Sustainability of Small Farms in Coastal Ecosystem

NIPA® GENX ELECTRONIC RESOURCES & SOLUTIONS P. LTD.
New Delhi-110 034

Sustainability of Small Farms in Coastal Ecosystem

Chakradhar Satapathy
Bishnupriya Mishra
Rabinarayan Mishra

NIPA® GENX ELECTRONIC RESOURCES & SOLUTIONS P. LTD.
New Delhi-110 034

NIPA® GENX ELECTRONIC RESOURCES & SOLUTIONS P. LTD.

101,103, Vikas Surya Plaza, CU Block
L.S.C. Market, Pitam Pura, New Delhi-110 034
Ph : +91 11 27341616, 27341717, 27341718
E-mail: newindiapublishingagency@gmail.com
web: www.nipabooks.com

For customer assistance, please contact
Phone: + 91-11-27 34 17 17
Fax: + 91-11- 27 34 16 16
E-Mail: feedbacks@nipabooks.com

ISBN: 978-93-58870-12-1

Composed and Designed by NIPA®.

Preface

Problem of agriculture in India is the problem of small, marginal and land less farmers. Our farm scientists have evolved adequate location specific technologies to make the farming occupation profitable. The increase in population and growing demand for quality food, clothing, shelter, fuel and fodder have imposed challenges before us. It is more so far rural areas where about 70 per cent people reside and work in different kinds of livelihood systems. The answer to these growing demands is sustainability of small farms that support majority of farm families. According to Dr. Swaminathan evergreen revolution would be triggered by farming system that can help produce more than the available land, water and labour resources without ecological and social harm. The authors have tried to analyse different dimensions of small farm units, agro-economic scenario, small farm production systems, organic farming, perception of small farm operators, adoption of technology, extension approach and strategy focusing attention on Odisha as unit of research dimensions. The book contains enough inputs for research and extension workers to look into small farms and their owners in view of changing scenario of our farming system. However, the book would serve as foundation to provide encouragement to the scientists and extension researchers to focus more on small farm units. The authors would thank Dr. Manoranjan Kar, the Honourable Vice Chancellor of Orissa University of Agriculture and Technology for his advice, encouragement and well wishes in writing the book.

Authors

Contents

1

Introduction

With increase in population, demand for basic needs has been steeply rising during the past decades in most of the developing countries of the world. The growing population needs food, clothing, shelter, fuel and fodder for them and their livestock. In India, 50 to 70% of the people are living in rural areas that neither have adequate land holdings nor alternate service opportunities to produce or procure these basic commodities.

The present day living demands quality life. The quality life can be maintained provided our surroundings remain intact or improved to provide means for better living. With increasing advancements in science and technology, we have manipulated our environment without paying adequate attention to consequences. The result has become so alarming that now all the countries of the world are trying to actualize the meaning and concept of sustainability. It is more so in case of agriculture as it forms core base of life supporting system.

The magnitude of manipulation varies from country to country depending upon level of development. It goes without saying that development and manipulation of environment are interrelated having positive and negative effects. To eradicate ill effects of manipulation, we need to understand the meaning of actualizing of sustainability.

To illustrate ill effects of manipulation it is observed that about one third of world's cropland is losing top-soil leading to long-term effect on productivity. About 50% of world's rangeland is overgrazed leading to formation of desert. Worldwide, almost half of the crop grown land are degraded at low to moderate

rate. The fodder needs of livestock in all the countries of the world now exceeded the sustainable yield of rangelands and other forage resources. The fodder demand in India is estimated to be seven thousand million tons against which sustainable supply is around fifty percent of the requirement. According to an estimate, about 30% of world's cropland is now loosing top soils at a rate that is undermining its productivity leading to food scarcity. If this trend continues, the world as a whole may face a dangerous situation, as it has been observed in country of Kazakhstan in Asia, Europe and Kenya and Somalia in Africa. It is surprising that the cropland area in coastal provinces of China because of top soil erosion and construction of factories and roads are going out of cultivation. Yield losses due to soil erosion records about 80 million tones in African countries that need attention of the scientists.

Water scarcity is another ill effect of present system of development. Water scarcity is increasing whereas in each year less than 80 million additional people are striking their claims to the earth's water resources. The manifestation of this problem is observed in the drying of rivers in several countries of the world. The Amu Darya in Central Asia, the Yellow river in China, the Nile in Egypt, the Jordan in Israel, the Ganges in India and the Mekong in Indo-China have become victims of this problem. It is further coupled with falling of water tables due to over pumping. Northern China is drying for depletion of water leading to the closing of wells, which serve for drinking purpose. In Punjab, the food basket of India, the water table is falling by 0.6 m per year, forcing farmers to drill deeper and deeper. In the South east plains of USA, practicing irrigated agriculture with water pumped from the *Ogallala aquifer* with little recharge, the farmers are forced to return to dry land farming.

According to FAO, 270 million hectares of land in the world are irrigated and these consume 70% of world's total human induced water consumption. About half of the water used for agriculture in USA comes from underground, from which it is pumped to the surface. The same is the situation in Indo-Gangetic plains of India. The water depletion gathers more momentum leading to slow ground water recharge and fall in water table. Over 80% of the captured water is used for irrigation in India, 13% to 15% goes to energy and industrial use and 4% to 5% is consumed for domestic purpose including that needed by livestock: how ever India may become a moisture deficit state for the water for irrigation if the present state continues (Katyal, 1997).

Forest is another important component of environment. It is reported that half of the forest are sinking due to exploitation of forest resources. The earth's forest area, estimated to be 5 billion hectares at the beginning of twentieth century has now reduced to 2.9 billion hectares. This forest loss is concentrated in developing countries. The ill effects of deforestation experienced disastrous flooding, alteration in rainfall cycle, reduction in natural vegetation and other natural calamities. The environmental scientists are in opinion that same trends on environmental degradation will reinforce each other showing synergetic effect. Such synergetic trends are described as often irreversible (Lester Brown, 2001).

Not only soil, water and forest but also fishery component has been seriously affected. An estimate shows two third of oceanic fisheries are now being fished at or beyond their capacity. With excessive demand on fisheries, the world is witnessing collapse of fisheries all over the inland, coastal and oceanic fisheries and commercial fishing is under great threat. Damage to coral reefs, a breeding ground for fish in tropical and sub-tropical waters is also taking a toll.

Another major threat concerns to the scientists is disappearance of the species. The archaeological records show five great extinctions since life begun in the earth. The world is now in the early stage of sixth great extinctions. Unlike the previous ones, which are caused by natural phenomena, this one will be of human origin. The World Conservation Union (IUCN, 1975) has pointed out the large increase in the critically endangered species in all categories – bird species, mammal species, fish species, amphibians etc.

Over the last three decades global average temperature has risen from 13.99°C in 1969 – 1971 to 14.43°C during 1998 – 2000. If carbon dioxide concentration in atmosphere doubles pre-industrial level by the end of the century reaching 560 ppm, the temperature will rise by 1.4°C to 5.8°C. This rise will lead to more extreme climatic events like heat waves, melting of ice, rising sea level and more destructive storms. In the Arctic Ocean, sea ice is melting. This may lead to expansion of ocean at the expense of continents with consequences such as inundation of coastal areas, salt-water inclusion and beach erosion.

The word 'Sustainability' is of recent importance with increasing consciousness among the scientists / planners and common people. Though the dictionary meaning of sustainability is 'keeping on efforts going continuously', 'ability to last out and keep from falling'; sustainability with reference to agriculture refers to the capacity to remain productive while maintaining the resource base. Sustainable agriculture is the successful management of resources for satisfying changing human needs while maintaining or enhancing the quality of environment and conserving natural resources (TAC/ CGIAR, 1989).

Sustainability in agriculture is thus considered from another important considerations of non-stagnant productivity in addition to environmental conservation and non-stagnant productivity. The non-stagnant productivity encompasses food grain availability and trend in agro-ecosystem consisting of irrigated area, coastal belt, rainfall area, hill and mountain areas. To have sustainable agriculture many key elements have been identified which enlists zero tillage, crop waste management, organic matter recycling, soil enrichment, integrated pest management (IPM), use of information technology and precision farming. To be very specific, agriculture is only sustainable if it ensures that today's development is not at the cost of tomorrow's prospects (World commission on economic development, 1987). Further, the commission also pointed out that major component of sustainable agriculture is economic profitability and environmental health. The round table on environment economy (1994) identified key acts of agricultural sustainability as (a) the impact on rural economy system within its capacity, (b) present and projected environment, (c) depletion of non-renewable resources, (d) diversity and innovated resource efficiency, (e) equitable distribution of benefits and (f) participatory productive systems, area of information and education for sustainable agriculture. To be more specific the eminent Agricultural scientist Swaminathan (1991) has enlisted as many as 15 parameters like Technological Appropriability, Economic Feasibility, Economic Viability, Environmental Soundness, Temporal Stability, Resource Use Efficiency, Local Adaptability, Social Acceptability, Social Sustainability, Political Tactness, Administrative Manageability, Cultural Desirability, Renewability, Equity and Productivity to define the operational aspect of sustainability.

Irrigation of salt rich arid and semi-arid lands has added environmental cost. When water use is excessive, the underground water table rises and brings with it the dissolved salts from the lower region of soil to the surface. The process resulted in low crop yield, increasing evaporation, leading salt-coated surface on crop land. Estimates suggest that in Bangladesh, river delta salinization is significantly reducing the yields in about one-third of crop land, while in the desert of south west US the figure approaches one-fourth. The coastal ecosystem of Indian sub continent is equally vulnerable to such type of hazards also.

The above scenario leads to conclude that manipulation of environment for development in agriculture has brought ill effects in all the components that constitute/ correlates agriculture. The manipulations can be arrested provided people gain sufficient insight to the problems and make use of recommendations to preserve the same. The state of Odisha in the map of Indian agriculture is not exceptional to the problems emerging out of environmental degradation resulting out of development patterns mostly undermining environment. The state of Odisha hence being a costal state of the country represents the all most all type of environmental hazards that pose threat to the agricultural sustainability with specific reference to the small and marginal farm families constituting major force in the field of agriculture. With this back drop the state Odisha was preferred to be selected in order to undertake a study on "Sustainability of small farms in coastal ecosystem" with a view to assessing the sustainability index considering various dimensions of sustainability of agriculture.

Odisha comprises of 4.47% of India's land-mass and accounts for 3.58% of population of the country. Nearly 85% of the population living in rural areas, depend upon agriculture for livelihood. According to estimate, a total of ground water resource in Odisha is 16,68,914 hectare meter. The gross annual draft for all use was 4,36,202 hectare meter. Therefore about 26.14% of ground water resources have been used in Odisha so far. The total cultivable land of the state is about 64.09 lakh hectare of which 35.03% have been provided with irrigation facilities covering 40% of cultivable land. The state is divided into 10 agro-climatic zones on the basis of soil, weather and other characteristics. The total cultivated land of Odisha is 61.80 lakh ha out of which 29.14 lakh ha (47%) is high land, 17.55 lakh ha (28%) is medium land and 15.11 lakh ha (25%) is low land with different texture and structure of soil.

The state's economy is greatly influenced by agriculture, which contributes about 17.49% to the Net State Domestic Product (NSDP) and 15.35% to Gross State Domestic Product (GSDP) of Odisha in 2011-12. With 46.67 lakh operational holdings, the small and marginal farmers account for 91.86% and about 70.34% of operated land is owned by small and marginal farmers (2010-11). In the field of irrigation, forestry, fishery and animal resources, the position of state is now in an improvement side. To support the development of the state, there is an urgent need to adopt recommendation for sustainable agriculture. The present problem of agriculture in coastal area is salinity, which comes to be 2,55,121 ha. The water logging areas (0.60 lakh ha) of the state is no less important to bring the attention of the policy makers for immediate solution. Out of the undivided 13 districts, Cuttack and Puri are the most affected area & having as much as 21.36 and 13.95 thousand hectares under waterlogged condition respectively. The total waterlogged area of the state is reported to be 74.76 thousand hectares. The acid soils of the state covering laterites and lateritic soil have abundance of iron that poses problems in agriculture. Highly acidic and alkaline soils limit crop growth and should be ameliorated for optimum crop production. In Odisha high acidic lateritic soils are observed in Cuttack, Puri, part of Balasore, Ganjam and Dhenkanal (pH 4.0 to 6.5). Sand dunes in coastal Odisha are the other important problems due to strong prevalence of wind. The dunned lands have poorest soil from agriculture point of view. The undivided districts like Cuttack, Balasore, Puri and Ganjam had sand-dune areas to an extent of 56298.40 ha. Strip land along roadside as well as rail line can be used for growth of agriculture of the state.

The above-mentioned situation of the state needs urgent planning for sustainable agriculture. The farmers of Odisha are required to have insight into problem so that attempts of the Government can bear the success whenever sustainable measures are put into action. Keeping above considerations in view, the present study was designed to understand the status of agriculture sustainability in Odisha in general and the factors affecting sustainability of agriculture under small farm production system in coastal ecosystem of Odisha with the following objectives in particular.

Objectives

1. To study the socio-economic profile and farming related factors of the small operators covered under study.
2. To measure the perception of small farm operators about elements of sustainability of agriculture and gap in comprehending sustainability.
3. To determine the relative position of study area under Sustainable Agriculture Index (SAI).
4. To determine the relative position of sub-components of factors of sustainability of agriculture as perceived by small farm operators.
5. To study the awareness and adoption of organic farming practices for sustainability of agriculture.
6. To study the agribusiness opportunities for sustainability of agriculture as perceived by the farmers under study.
7. To find out effective extension approaches for enhancing sustainability of agriculture to provide better living standard to farmers at large.
8. To find out suggestions and formulate strategy for preservation and maintenance of sustainability of agriculture.

Scope and importance of the study

The study makes an effort to analyze socio-economic and farming related factors of the small farm operators. It also investigated into the perception of the sample about sustainability of agriculture on five important parameters like, ecological soundness, economic viability, social accessibility, humaneness and adaptability. It also determined existing gap comprehending the sustainability of agriculture, Sustainable Agriculture Index (SAI) of the area as well as the relative position of sub-components of factors of sustainability as per reaction of the sample farmers under study. The scope of organic farming and agribusiness opportunities towards sustainability of agriculture was studied in farmers perspective. There are recommended measures of sustainability of agriculture. In this regard the adoption of recommended measures, extension elements as well as suggestions for sustainability of agriculture have been analyzed with appropriate statistical tools. The findings of the study would form a base for development agencies in formulation and implementation of the programmes for sustainability of agriculture in the state. The investigation has been designed to find out effective extension approaches. The suggestions regarding preservation and maintenance of sustainability have been recommended as per the perception of experts and farmers. Thus, the information generated through the study adds to the already available recommendations for sustainability issue in field of agriculture and allied sectors in the state. The assumption is that the results of the study would be useful to the planners, executants and field functionaries in designing appropriate program and effective implementation of the programs on sustainability of agriculture in the state.

2

Dimensions of Sustainable Agriculture

Concept of sustainable agriculture is of multiple dimensions. It denotes potential of technology, fragile ecosystem, and human interference of limited natural resources. An understanding of sustainable agriculture in brief indicates:

1. Increase in production with efficient use of resources
2. Control of external inputs of synthetic nature
3. Nutrient cycle within the farm must be closed.

Technical Advisory Committee (TAC) of CGIAR reveals that sustainability is a dynamic concept that allows for changing needs of a steady increasing global population. Sustainable agriculture should involve the successful management of resources for agriculture to satisfy changing human needs while maintaining or enhancing the quality of environment and conserving natural resources.

The dictionary meaning of sustainability is to keep an effort continuously the ability to last out and keep from falling. Sustainability basically refers to the capacity to remain productive while maintaining the resource base. Locker (1988) explained the term sustainable, a time dimension and the capacity of farming system to endure indefinitely. The concept of sustainability should serve as guide to agricultural practices. It must include technology and practices that both sustain and enhance productivity to meet the increasing food demands

(Rutton, 1988). In short, environmental sustainability implies that it is compatible with and supportive of the criteria like;

i. Meeting basic needs of the people
ii. Keeping population density below carrying capacity of the region
iii. Adjustment to consumption pattern of the population
iv. Conserving, recycling and keeping priority for use of nonrenewable energy
v. Sustainable agriculture is a way of practicing agriculture which seeks to optimize the skill and technology, to achieve long term stability of the agriculture enterprise, environment protection and conserve safety.

The study entitled "Sustainability of small farms in coastal ecosystem" is of multidisciplinary in nature. Both social science and field related sciences are merged together to present views about sustainability of agriculture in general and ecological aspects in particular. In course of mapping of literature from different disciplines, a number of variables were examined. The relevance of these literatures and theme of the present study was critically examined. The available relevant literatures were grouped under the following headings.

Reviews on Dimensions of Agriculture

2.1. Concept, meaning and definition of Sustainable Agriculture (SA) - A review

According to Gips (1986) sustainable agriculture is one, which is ecologically sound, economically viable, socially just, humane and adaptable.

Brundtland Commission (1987) opined that sustainability essentially is a holistic notion at which one can exploit exhaustible resources or the level of degradation of renewable resources that may be permitted.

According to the "International Green Front Report" (1988), "sustainable agriculture is a broad term which includes organic farming, culture, bio-dynamic farming and other forms of progressive agriculture where by the farm and surrounding ecosystems are enriched rather than impoverished.

TAC/ CGIAR (1989) reported that sustainable agriculture is the successful management of resources for agriculture to satisfy changing human needs while maintaining or enhancing the quality of environment and conserving natural resources.

Reetz (1990) stated that sustainable agriculture is the system that is profitable and environmentally sound. Sustainable and conventional agriculture can be one and the same.

Board for International Food and Agricultural Development and Economic Co-operation (BIFADEC, 1991) opined "Sustainable agriculture is not a technique to be precisely defined, but a concept common to most environmentally based economic activities. Simply, it is that, continued production depends upon maintenance of the resources on which production is based".

According to Swaminathan (1991), Sustainable agriculture has following dimensions like Technological appropriability (how far the agriculture technology be it a seed, fertilizer, pesticide or improved machinery suits the social and infrastructural situation of the end users), Economic feasibility (the capacity of farmers to afford to incorporate the technology in his farm within his realm of financial status and position), Economic viability (the return to investment of every rupee counts), Environmental soundness (whether technology results in enriching the environment or at least doesn't harm the existing agro-ecological conditions), Temporal stability (whether the positive aspect remain stable over the long run), Resource use efficiency (how efficiently the technology can utilize the inputs to convert them into useful, productive and eco-friendly output), Local adaptability (the extent to which the technology is adaptable to the existing local condition of the farmers), Social acceptability (the extent to which the technology is acceptable by different sector of the society), Social sustainability (has to meet the personal needs), Political tactness (whether the technology can be used unharmed in the existing intricacies and implications of political setup), Administrative manageability (the extent to which the technology can be practically implemented under the existing bureaucratic structure), Cultural desirability (the extent of the technology fits with the culture patterns and values of the society), Renewability (the extent to which the technology can be used or reused without much additional efforts or inputs), Equity (how evenly the product of agro-ecosystem are distributed among the local producers and consumers) and Productivity (the rate and amount of production per unit of land or input). It also measures sustainability of agriculture.

Venkatramani (1991) revealed that sustainable agriculture is that form of farming which produces sufficient food to meet the needs of the present generation without eroding the ecological assets and the productivity of the life supporting system of future generations.

Harrington (1992) pointed out that sustainable agricultural development could be interpreted in terms of agro-ecology, equity and sustainable growth.

Izac *et al.*(1992) reported that sustainable agriculture contains three equally important components namely environmental quality and ecological soundness, plant and animal productivity and socio-economic viability indicating the quality of life and profitability for farmers, farm workers and rural communities.

Katyal *et al.* (1992) pointed out that sustainable agriculture revolves around the use of low cost inputs, less dependence on chemical fertilizers, maintenance of soil fertility and ecological harmony, preservation and enhancement of biological productivity.

Lefroy and Hobbs (1992) revealed that water cycle, nutrient cycle, energy and diversity are the ecological parameters for the assessment of land use on farms in the wheat belt of Western Australia.

Reijntjes *et al.* (1992) opined that agriculture is sustainable if it is ecologically sound, economically viable, socially just humane and adaptable. Further, they opined that objectives of any household are productive (output per cent of land and input used), security (minimizing risk of production), continuity maintaining soil and water health) and identity (self respect, social justness and humaneness). Feeling of identity is maintained by technologies that permit to be self reliant and to control decision making about use of local resources and products.

Nadakarni (1993) observed that the sustainable development has observable indicators like (i) productivity (yield of net income), (ii) stability of yield or net income, (iii) sustainability of yield and net income and (iv) equitability in terms of income distribution.

Mendoza (1994) expressed that any agricultural system to be identified as sustainable one, must achieve the objectives like, increased productivity, profitability and economic well being of farmers, preservation of environmental quality, self sufficiency of food, equitable distribution of economic returns and optimal utilization of internal resources in farm.

Stockle *et al.* (1994) concluded that all definitions of sustainable agriculture include food productivity, safety, resource protection, quality of life and environmental quality specifically emphasizing on attributes like profitability, productivity, soil, water and air quality, energy efficiency, fish and wild life habitat, quality of life and social acceptance.

Chakraborty *et al.* (1995) remarked that the essence of sustainable technology is one which is economically viable, dynamic in nature, adaptable, acceptable to farmers, prevent degradation or excessive loss of soil, water, nutrients and conserve resources for future generation, avoid over exploitation of land with genetic diversity and ecological integrity ensuring good health and vitality of people.

Katar (1995) found that sustainability is defined as the ability of a renewable natural resource system to produce a socially optimum level of output in perpetuity with no detrimental effects on the resource system itself and on the physical environment.

Gowda (1996) indicated that nutrient management, water management, integrated pest management, crop yield security, information, self reliance, input self sufficiency, family food self sufficiency, land productivity and input productivity are some of the important indicators of sustainability of agriculture.

Vijaya (1996) opined that sustainable agriculture can be referred to as alternate farming or conservation farming or regenerative agriculture or organic farming for accepting the principle of co-existence of soil microbes livestock and free component which leads to natural farming for future prosperity.

Yadav *et al.* (1996) found that the harmony between increasing production through advanced agro technology for the present and preservation of environmental quality and resource base for the future led to the concept of sustainable agriculture. It aims at stability of production overtime through the location and situation specific production principles.

Agarwal (1997) stated that eco-sustainable agriculture is linked to carrying capacity of the ecosystem based on the development planning process by analyzing the underlying correlationship between, population, poverty and pollution to provide supportive capacity for development and assimilative capacity for maintenance of acceptable quality of environment.

Lahiry (1997) stated that sustainable development is a multidimensional concept with three interacting angles of ecology, economics and ethics, the necessary conditions for achieving sustainable development, ecological security, economic ability and social equity. Sustainable agriculture calls for an environmental friendly technology at the global level and agreement among the nations to respect each other's need.

Nagabhusanam (1997) reported that crop diversity, land productivity, input use index, integrated pest management, eco-system management, soil environment level, crop productivity, enterprise supporting ability, social equity, and carrying capacity are some of the important indicators of sustainability of agriculture.

Sadangi (1997) found that the adoption of proven alternative system would result on even greater economic benefits to farmers and environmental gains for the state and the economy.

Satapathy and Das (1997) stated that specific pin pointed stimulating extension methods which can be used as per the clients bring sustainability in agriculture, the main drive to maintain good human being along with good farmers. This will help us to create eco-friendly situation leading to sustainable agriculture, which will encompass stability, productivity and equity.

Verma and Agarwal (1997) reported that economics of sustainable development is the development that lasts. It means the need of the present without compromising the ability of future generations to meet their needs.

Teng (1999) stated that sustainable agricultural development must satisfy the criteria of economic viability, ecological soundness and social equity and would generally have to be practiced in the context of conservation of biodiversity, the balancing of present with future needs, a renewal of the natural resource base and the augmentation of indigenous knowledge.

Swaminathan (2000) emphasized on the major components of integrated natural resource management of integrated gene management, integrated nutrient management / supply, efficient water management, soil health care, integrated pest management and efficient post harvest management.

Fadini and Lauzada (2001) stated factors determining biodiversity in agro-ecosystem. The importance is emphasized on replacing harmful agricultural practices with less damaging practices as a way of achieving sustainable agriculture.

Hasbi (2001) opined three important points in sustainable agriculture i.e. the cultivation system, organic and natural fertilizers, microorganisms and erosion control.

Jitsenguan (2001) described five sustainable farming systems for small-scale farmers and their profitability, i.e. (i) integrated farming, (ii) organic farming, (iii) natural farming, (iv) agro-forestry and (v) new theory farming. All these were promoted as alternative agriculture to small Thai farmers. It was shown that sustainable agriculture was not only financially viable but also enhanced environmental quality for the benefit of farmers.

Rai *et al.* (2001) stated that management practices for sustainable agriculture differs from those of modern agriculture and recommended important steps for suitable agriculture like (i) watershed management, (ii) integrated development and conservation of genetic resources, (iii) agro-forestry, (iv) integrated nutrient management, (v) tillage, (vi) ground water management, (vii) remote sensing in management of natural resources and (viii) integrated pest management.

Sharma (2003) opined that simplified processes sustained for long time while more dependence on external inputs (exotic plants, seeds, nutrients, irrigation etc.) increase cost and reduce sustainability.

Sabarathnam (2002) emphasized five factors for sustainable agriculture as Ecological soundness, economic viability, social accessibility, humanness and adaptability.

Rai (2005) remarked that sustainable development of agriculture in the 21st century depends largely on the emerging technologies in agriculture and found encouraging. Agricultural fields vary with space and time and hence required site specific crop management for maximizing the efficiency of the input resources.

2.2. Knowledge, perception and attitude about sustainability of agriculture

Nirmala *et al.* (1995) stated that nearly half (42.83%) of the farmers were found to possess a high level of knowledge followed by medium (32.50%) and low (21.67%) knowledge on bio-fertilizers.

Gowda (1996) stated that most of the farmers (67.77%) possessed a favorable attitude towards sustainable farming.

Padmavathi (1996) observed that 65.50% of Mitrakissans had favorable attitude followed by 18.88% and 15.56% with high and less favourable attitude towards National Watershed Development Project for Rainfed Area (NWDPRA).

Reddy (1996) found that majority (55.83%) of the vegetable growers had medium knowledge followed by high (26.65%) and low (17.29%) levels of knowledge respectively.

Jirli and Gangadharappa (1997) stated that nearly half (45.00%) of the farmers were found to posses a medium level of knowledge followed by high (31.00%) and low (20.00%) knowledge level respectively with respect to integrated pest management of vegetable cultivation.

Rambabu (1997) indicated that majority of the respondents (70.10%) had favorable attitude towards indigenous technology items in wet cropping system followed by neutral attitude (13.34%), unfavourable attitude (8.33%), highly favorable attitude (5.10%) and highly unfavorable attitude (3.33%) in wet cropping systen. In dry cropping system majority of respondent (66.67%) had favourable attitude towards indigenous technology items followed by unfavourable attitude (11.67%) and neutral attitude (5.00%) respectively. On an average majority of the respondents (73.34%) had favourable attitude towards indigenous technology items followed by 10.00% each of neutral and unfavourable attitude, and 3.33% each of highly favourable and highly unfavourable attitude towards the practice.

Singh (1997) reported that 55.00% of respondents had a favourable attitude and 5% a highly favourable attitude towards the social forestry programme.

Bhople and Lakhdive (1998) revealed that more than 50% of the farmers knew the IMP practices.

Manjunath *et al.* (1998) indicated that majority of the farmers had perceived the practices of green manure as simple, not risky, useful but costly and not practicable.

Srivastava (1998) opined that majority of the farmers had a favourable attitude towards improved farm implements.

Alagesan (1999) stated that about two third (61.67%) of the respondents had knowledge about azospirilum, followed by rhizobium. The knowledge on other bio-fertilizers by the farmers was less than 5.0%. Farmers have limited knowledge on parasite and predators ranging from 0.83% to 9.13% with respect to bio-fungicides. Only 6.67% of farmers had knowledge on *Trichoderma viridae.*

Haque *et al.* (1999) informed that 52.00% of farmers had a medium level of knowledge followed by 37.00% had a high knowledge level and 11.00% had a low knowledge level.

Khan (1999) indicated that more than three fourth (77.77%) of the respondents were having medium knowledge on eco-friendly technologies in rice cultivation while 13.34% and 8.89% had high and low knowledge respectively.

Kumar *et al.* (1999) indicated that many of the farmers were neutral or undecided in their attitudes towards agro-forestry programme.

Nagdeve (1999) informed that majority (68.67%) of the respondents had moderately favourable attitude, while 16.66% had more favourable attitude and 14.67% respondents had less favourable attitude towards IPM in rice crop.

Ohlander *et al.* (1999) indicated that future agricultural system should be aimed at goal oriented sustainable management of ecosystem.

Rahman and Mikani (1999) stated that half of the farmers in the study area possessed a less than favourable attitude towards sustainable issues of agricultural development. Only 6.00% of them expressed a highly favourable attitude and 44.00% expressed a moderate attitude in this regard.

Rahaman *et al.* (1999) revealed that most of the farmers possessed a high favourable attitude towards environment and sustainability issues of agricultural development.

Cothe and Barker (2000) stated that more than half (58.67%) of the respondents had medium level of knowledge followed by high (22.67%) and low (18.66%) knowledge level on bio-fertilizers.

Prasad and Sundaraswamy (2000) observed that 36.67% of the respondents belonged to less favourable attitude category followed by 34.33% belonging to more favourable category. A significant percentage (29.00%) of sample' belonged favourable category.

Resmy *et al.* (2000) observed that majority of small farmers and big farmers had medium level of knowledge. In case of small farmers it was followed by high level of knowledge and then low level of knowledge where as an equal percentage (25.00%) was seen to be distributed under high level of knowledge in case of big farmers.

Ramamurthy (2000) indicated that 72.50% of the respondents had favourable attitude followed by 11.67% with more favourable attitude and 15.83% with less favourable attitude towards Janmabhoomi Programme.

Raj and Satpathy (2003) found that the watershed beneficiaries agreed that, available resources, development of team work, good linkage with related departments and adoption of improved technology can increase production and productivity.

2.3. Ingredients that define the sustainability of agriculture

Altieri (1992) opined that key strategy in sustainable agriculture is to restore the agricultural diversity of the agricultural landscape. Diversity can be enhanced in time through crop rotation and sequences, cover crops, intercropping, agro-forestry and crop/ live-stock mixture without damaging the environment or depleting natural resources for future generations. Social and economic conditions in subsistence agriculture may preclude reliance upon costly or dangerous supplements such as expensive equipment, commercial fertilizers and pesticides.

Dunlap *et al.* (1992) reported nine dimensions viz. protect and enhance soil health, ensure supply of safe and whole food, improve site specific knowledge of farmers, enhanced environment and wild life habitat, increased diversity, improved farm economy, reduced agrochemical use, reduced energy use and reduced purchase of inputs being considered as most important in explaining the concept of sustainable agriculture.

Loker (1994) reported that low external input agro-forestry systems [(a specialized form of low input sustainable agriculture system of (LISA)] hold great promise as alternative sustainable production systems for small to medium farmers in humid tropics.

Jayashree and Prasad (1994) analysed different dimensions of sustainable agriculture and identified different dimensions which include technological appropriability, economic feasibility, environmental soundness, temporal stability, resource use efficiency, local adaptability, social accessibility, political tactness, administrative manageability, cultural credibility and renewability.

Lawrence (1994) proposed the indicators for sustainable agriculture as a guide to evaluate appropriate technologies i.e. (1) ecological indicators (2) economic indicators and (3) social indicators.

Cai and Smith (1994) opined that sustainable agriculture should combine all the components of sound husbandry into united packages.

Kutty (1996) reported the important dimensions of sustainable agriculture. Six dimensions adorned the top position. They are resource use efficiency, environmental soundness, economic viability, technological appropriability, economic feasibility and local adaptability.

Basawarajaiah and Patil (1997) reported that sustainable crop production not only involves identification and application of improved technology but also sustainable management of natural and renewable resources, maintaining the quality of environment, conserving the natural resources, soil and water on a long term basis.

Elangovan and Vasantakumar (1997) studied the usefulness index of different eco-friendly technologies as perceived by extension officers which includes IPM and bio-fertilizer application, FYM, green manure and neem cake application, compost application and integrated farming systems, forestation, neem seed kernel extract application, composted coir pith, vermin-compost, rice-fish-azolla culture.

Barad *et al.* (1998) indicated the improved crop production technologies, organic farming, alternative land use systems including agro-forestry systems, watershed management, cultivation of less economic plants, biological control, integrated pest management, utilization of wastelands, and use of biological wastes in agriculture as the vital ingredients of agricultural sustainability.

Jalali *et al.* (1998) reported the non target impact of pesticides, the parameters governing integrated disease management, the relevance of biological control, the biotechnology and disease management in reference to plant pathogens.

Koocheki *et al.* (1998) revealed that a wide range of criteria were evaluated, including climatic, edaphic, cultural practices, agro-biodiversity, extension and education, demographic factors, financial issues and marketing for having sustainability of agriculture.

Kern and Geiss (1999) recommended that main issues of sustainability in agriculture are protecting and improvement of agricultural resources i.e. preserving and restoring soil and water quality which is best met through integrated crop management practices.

Singh *et al.* (1999) reported that enhancement of genetic diversity, conservation tillage, improved level management, integrated pest management and integrated water-management systems were some of the important dimensions of sustainable agriculture.

Dabrowski (2000) stated that the recent shift from concept of Integrated Pest Management (IPM) as Intelligent Pesticide Management to ecologically based pest management to meet the long term goals of safe, economically viable and sustainable agriculture, would require development of new interaction between farmers, extensionists, researchers and policy makers.

Paroda (2000) indicated that to remain relevant and efficient there is a need for paradigm shift from commodity centered to an ecologically, economically and socially sustainable farming system centered approach to agricultural research (crops, livestock, fish, forestry and agro-forestry) and organising education, extension and skill empowerment on the basis of farming system intensification, diversification and value addition.

Anderson (2000) opined that past and current failure to replenish soil nutrient in low income countries must be rectified through the balance and efficient use of organic and inorganic plant nutrient and through improved soil management practices.

Swaminathan (2000) stated that the evergreen revolution would be triggered by farming systems that can help produce more than the available land, water and labour resources without their ecological or social harm.

Singh (2001) concluded that predominance of small scale agriculture necessitates a synergetic integration of crop, livestock and environment at farm level in the region to region as a long term strategy.

Al Jaloud *et al.* (2002) opined that there exists an excellent potential for a renewable crop production if certain soil water and crop management practices are considered in an arid environment.

Bhavsar (2002) opined certified organic food and fibre production as an important part of the agricultural economy.

Bussche (2002) indicated that in agriculture, sustainable development should however not be equated with organic farming. Given appropriate management, both farming systems organic as well as conventional farming can be ecologically and socially sustainable.

Cox *et al.* (2002) emphasized that high water use efficiency was needed to minimize the overuse of sea water resource.

Eichhorn *et al.* (2002) emphasized the importance of agricultural engineering in sustainable agriculture that involves the technical adoption for sustainable soil protection, energy cost sparing production procedures in crop production, pastoral management and care of the countryside including recycling of residual biological material.

According to Gelderman and Kogel (2002) solution of sustainable agriculture is more than ever a scientific development of tools for a sensible management of resources.

Jonson (2002) emphasized on the Integrated Production (IP) rules to be followed on entire farm state and crop protection should be based on preventive and cultural methods. Biological control methods should be given priority and the use of pesticides should be minimized through monitoring of pests.

Kassa *et al.* (2002) opined that any agricultural development initiative in Ethiopia must include both crop and livestock components for success.

Liu *et al.* (2002) found that after 20 years of soil conservation and restoration, the ecosystem soil health index of the watershed had improved from 0.178 in 1985 to 0.707 in 1999.

Yadav (2004) reported that in developing countries sustainable resources management, increased crop yields without much reliance on costly external inputs, environmental and biodiversity protection through organic farming, taking care of soil quality, water quality and air quality as well as energy use would maintain sustainability in crop production.

2.4. Adoption of recommended measures of sustainable agriculture

Muthuraman (1995) stated that 40.00% the farmers interviewed were medium adaptors and about 30.00% of each of them belonged to low and high adaptor categories in adoption of farm yard manure, with respect to green manuring 40.00% of farmers were in low, 44.00% of them in medium and only 16.00% of them were in high categories.

Nirmala *et al.* (1995) reported from their study that majority of the farmers (54.17%) were found in low level of overall adoption followed by one fourth (25.00%) in medium and one-fifth (20.83%) in high level of adoption of bio-fertilizers.

Khan (1999) indicated that majority (70.00%) of the respondents belonged to the medium category of adoption followed by low (15.50%) and high (14.44%)

category with respect to the adoption of eco-friendly technologies in rice cultivation.

Manjunath *et al.* (1999) observed that 36.00 % of the sugarcane growers had low adoption followed by 34.00% each were observed under high and medium adoption level of sustainable sugarcane farming practices of sugarcane.

Nagdeve (1999) reported that majority (68.00%) had medium level of adoption, while 16.67% had low adoption and 15.23% respondents had high adoption of IPM practices.

Sriram and Palaniswamy (1999) informed that 39.16% of the cotton growers belonged to medium level of adoption followed by high (35.84%) and low (25.00%) level of adoption of eco-friendly agricultural practices in cotton cultivation.

Cothe and Barker (2000) reported that majority of farmers had medium level of adoption followed by high (32.00%) and low (26.00%) level of adoption of bio-fertilizers.

Dhillon (2000) informed that majority of respondents had medium level of awareness and adoption regarding agricultural pollution in Punjab.

Saxena and Singh (2000) concluded that majority of the farmers belonged to medium level of adopter category who were following 5 to 7 out of 10 organic farming practices, whereas only 25.00% of the farmers were following more than seven kinds of organic farming practices.

Siddiq (2000) recommended that development and extensive adoption of location specific forming systems aiming at higher productivity, better return and year round employment would alone help improve the quality life of those inhabiting in the long neglected ecology.

Sumathi and Alagesan (2000) informed that nearly little less than half (41.00%) of the farmers had medium level of adoption followed by 35.00 percent at low level and 25.00% at high level of adoption.

Bhople *et al.* (2001) found that majority of the respondents (81.33%) had low level of adoption of bio-control practices for pest management in cotton followed by medium (16.67%) and high (2.00%) level of adoption.

Kadam *et al.* (2001) observed that majority (68.00%) of the beneficiaries had adopted only one practice, namely, 'dividing fields with small bunds'. The practices namely, 'Stubble and agro-waste plucking' (38.66%), and small earthen bunds (23.33%) were also adopted by a considerable number of the beneficiaries.

Kalaskar *et al.* (2001) revealed that majority of the respondents (67.27%) were moderately aware of different IPM practices in cotton. In other words it could be stated that quite a few (16.97%) respondents have high level of knowledge about IPM practices.

Cox *et al.* (2002) stated that to maximise the adoption by farmers, water use efficiency must be based on easily measured parameters.

Bautista (2003) informed that the introduction and adoption of mechanised system should also consider selective mechanisation for suitable operations, socio-economic factors, as well as environmental consensus.

Singh and Singh (2004) opined that sustainable development has become key word in every policy of Government. But in present population explosion scenario it is important to provide alternative means to the ever increasing population otherwise all policies will be eye wash only.

2.5. Extension and communication approaches for enhancing sustainability of agriculture

Padhee (1995) reported that non-availability of required inputs, lack of extension efforts and methodology directed towards small farmers, restricted access to institutional credit facilities, and the suspicions of farmers regarding IPM recommendations were major problems in sustainable pest managment.

Reddy (1996) indicated that lack of resources, lack of follow up actions by implementing officials, lack of technical guidance by scientists, lack of efforts on the part of implementing agency to educate and convince the farmers about the importance of people's participation in programme, lack of training facilities to all farmers, most of the watershed technologies requiring high level of management, lack of adequate contacts with people in pre-project and planning stages, officials contact with few farmers only in the village and lack of cooperative effort among the people were the problems expressed by the farmers participating in watershed development programme.

Rambabu (1997) indicated that lack of more supporting price to the quality products was the major constraints in wet cropping system followed by lack of good grading facilities in dry and garden cropping system. Lack of publications on proven indigenous technologies, lack of proper training facilities and lack of organized extension machinery to disseminate the proven indigenous technologies were also the major constraints in wet, dry and garden cropping systems respectively.

Pornthip (1998) reported that the people's actual participation and involvement in the activity of community forest management was at low level.

Lakhan *et al.* (1999) reported that the major constraints identified were lack of trained staff, time limitation, lack of co-operation from villagers at initial stage while formulating village development plan, shortage of staff and excessive work load, frequent interference by community representatives in project and lack of faith among villagers, quick planning and poor rapport, while eliciting peoples participation.

Nagdeve (1999) revealed that in respect of socio-economic constraints, the most important constraints were: lack of credit facilities, high cost of inputs, and higher interest on credits. With regard to communication constraints, most important constraints were lack of precise information from change agents and difficulty to contact the extension agency at the time of need.

2.6. Suggestions for maintenance and preservation of sustainability of agriculture

Padhee (1995) reported that education of farmers, IPM research, effective extension network and government policy were essential for development of agriculture.

Reddy (1996) opined that forming user and self-help groups in each village, making proper follow up actions by officials, local communications, institutions and leaders must be encouraged to involve in analyzing the problems and interests of people, implementing agency should intensify its extension activities to motivate the farmers' involvement in programme, timely supply of critical inputs and implements, giving equal representation to all categories of farmers in watershed society, provision of timely technical guidance by scientists, provide credit facilities to the needy farmers, giving much emphasis on horticulture and income generating activities in programme, organizing training programmes to all beneficiaries on all components of programme, watershed officials should support indigenous technologies, constructing more number of farm ponds and percolation tanks in these areas, and entrusting the development of community lands to village institutions were the suggestions given by the farmers to improve their participation in watershed development programme.

Rambabu (1997) revealed that the important suggestions made by the farmers to overcome the constraints were improvement of sound research on ecological farming practices making available the bio-fertilizers and neem products on subsidized rates by government, establishment of bio-gas plants for getting more organic manure and strengthening the extension system for immediate dissemination of proven indigenous technologies.

Shinde *et al.* (1997) inferred that cent percent of cotton growers suggested for provision of certified seeds of cotton to prevent initial pest infestation, followed by more than 90.00% of them suggesting to provide a kit comprising magnifying lens to observe the eggs and larvae on the leaves, literatures having details about the IPM practices, pheromone traps to monitor the pest occurrence, facilitating the control through neem seed and making available the neem seed kernel to promote its use against the boll-worm as the important recommendations in IPM programme.

Nagdeve (1999) revealed that the important suggestions made by the IPM trained farmers to overcome the constraints were timely supply of inputs at reasonable cost, regular farm visit of extension workers, arranging more credit facilities, less rate of interest on loan, providing the subsidy on inputs. Timely technical guidance, organizing more number of training programmes and follow up of training programmes for successful implementation of IPM programme and thereby improving the plant protection status of the farmers were also suggested.

Pornthip (1998) recommended that the key persons of the various groups should be educated so that they would perceive the importance of establishing the organization to take the charge of looking after village's forest expecting that these key persons would further lead their group members to have greater awareness and participation in the project activity for sustained forest management.

Raju and Radhakrishna (2000) observed that the respondents suggested ways to overcome the constraints in betel vine cultivation by exact control measures of diseases and pests, continuous power supply, timely technical guidance, frequent monitoring of technical staff, maintaining demonstration plots, provision of fertilizers on subsidy, release of cyclone resistant varieties, conducting training programmes, increase the price of farm produce, supply of effective plant protection chemicals, providing bank loans, establishing soil testing laboratories and collection centers nearby in the order for benefit of betelvine growers.

Satapathy and Das (1997) reported that specific pinpointed stimulating extension methods, which could be used as per the clients bringing sustainability in agriculture. The main drive is for development of good human beings along with good farmers. This would help us to create eco-friendly situation leading to sustainability in agriculture, which would encompass stability, productivity and equity.

Swaminathan (2000) opined that the future would belong to small farm families taking to precision agriculture, involving the use of right inputs at the right time and in the right way. He also suggested that eco-technology based precision farming could help cut costs, enhance marketable surplus and eliminate geographical risks. This would be the pathway to an evergreen revolution in small farm agriculture.

Flora (2001) indicated that attention to differences within the communities was important in working towards sustainability of an agro-ecosystem.

Efforts to build social capital among women proved critical in developing both collective and household strategies for sustainability.

Singh (2004) recommended that the use of insecticides, fungicides and herbicides of plant origin was needed and their extract dose to be recommended. This would go long way in providing semblances and equilibrium in the agricultural ecosystem and environment as a whole.

Summarizing available literatures on sustainable ingredients that sustain the farm production system, the following groups can be formed.

Table 1: Indicators of different groups of sustainable farm production system

Group		Indicators
1. Crop and Cropping system	1.	Crop rotation
	2.	Cover crop
	3.	Inter cropping
	4.	Diversity in farming
	5.	Farming system intensification
	6.	Rice-fish-azolla culture
	7.	Integration of crop, livestock and environment
2. Mixed farming approach	1.	Agro-forestry
	2.	Crop and livestock
	3.	Alternate production system
	4.	Sound husbandry
	5.	Integrated farming system
	6.	Forerstation
3. Soil	1.	Soil health
	2.	Water and soil management
	3.	Alternate land user
	4.	Soil management
	5.	Soil protection
	6.	Recycling of biological residual materials
	7.	Soil Conservation
4. Environment	1.	Environment
	2.	Wild life
	3.	Ecological indicators
	4.	Natural renewable resource
	5.	Use of waste land

(Contd.)

Group		Indicators
	6.	Biological wastes
5. Input use	1.	Reduction in use of agro-chemicals
	2.	Reduction in use of chemical inputs
	3.	Integrated Pest Management (IPM)
	4.	Green manure
	5.	Compost
	6.	Vermicompost
	7.	Biological control
	8.	Integrated Disease Management (IDM)
6. Technology	1.	Technology appropriateness
	2.	Economic feasibility
	3.	Temporal stability
	4.	Resource use efficiency
	5.	Social acceptability
	6.	Cultural harmony
	7.	Local adaptability
	8.	Improved technology
	9.	Improved crop production technology
	10.	Organic farming
	11.	Balance use of plant nutrients
	12.	Value addition
	13.	Conventional farming
	14.	Bio-diversity
7. Farmer	1.	Training
	2.	Knowledge on sustainability
	3.	Farmer and scientists interaction
	4.	Skill improvement
	5.	Participation in policy making

The indicators of different groups of sustainable farm production system were essentially helpful to find out the factors affecting the sustainability of small farms and making insights to the factors responsible for posing problems in sustainable agriculture so far this small farm production system is considered.

3

Agro-Economic Scenario of Odisha

3.1. Brief description of the state

Odisha a coastal state of India lies in between 17^0 31' to 22^0 27' North latitude and 81^0 27' to 87^0 30' East longitude. It is bounded by West Bengal in the North East, Jharkhand in the North, Chhatisgarh in the West, Andhra Pradesh in the South and Bay of Bengal in the East. The main features of the state are given in Table 2.

Table 2: Odisha at a glance (2011 census)

Sl. No.	Facts	Unit	Magnitude
1.	Geographical situation	-	Between 17^0 31' to 22^0 27' North latitude and 81^0 27' to 87^0 30' East longitude
2.	Geographical area	Sq. km.	1,55,707
3.	Population	Number	41,947,358
4.	Male population	Number	21,201,678 (50.54%)
5.	Female population	Number	20,745,680 (49.46%)
6.	Density of population	Persons/sq.km.	269
7.	Literacy rate	Percentage	73.45
8.	Male literacy rate	Percentage	82.40
9.	Female literacy rate	Percentage	64.36
10.	Sex ratio	Female/ 1000 males	978

Source: Census of India 2011, Odisha Series 22, p.1.

3.2. Physiographic information of the state

Odisha is an extensive plateau, which slopes gently into coastal plain along the Bay of Bengal. The river Mahanadi is flowing west to east through the plateau dividing the state into two definite parts. The northern part is an extension of Chhotanagpur plateau and the southern part is a part of eastern ghat region.

The state has been divided into ten agro-climatic zones. The distinctive features of various zones are as follows.

i. North Western Plateau

The zone comprises of agricultural districts of Sundergarh, Baneigarh and Panposh in Sundergarh revenue district and Kuchinda under Sambalpur district. The climate is hot, moist and sub-humid having mean annual rainfall of 1185.5 mm, mean maximum temperature of 34°C and mean minimum temperature of 15.6°C. The soil is light textured with red and yellow soils. Major crops grown are rice, toria, wheat and vegetables.

ii. North Central Plateau

It consists of the agricultural districts Rairangpur, Baripada, Karanjia in Mayurbhanj district and Champua in Keonjhar district. The climate of the zone is hot, moist and sub-humid. The mean annual rainfall is 1268 mm with mean maximum and mean minimum temperature of 32.4°C and 20.6°C respectively. Red loam and light textured soils are mostly found in the zone. Rice, niger, maize, horse gram, black gram, arhar, sesamum and vegetables are the major crops grown.

iii. North Eastern Coastal Plain

Agricultural district Anandpur under Keonjhar district, Balasore and Bhadrakh districts are coming under this zone. The climate of the zone is moist and sub-humid with mean annual rainfall of 1600 mm. The mean maximum temperature is 33.2°C and mean minimum temperature is 22.6°C. Alluvial soils and narrow strip of sedimentary sandy loam to clay textured soils are the characteristics of this zone. Major crops grown are rice, jute, groundnut, toria, vegetables, black gram, green gram and betel vine.

iv. East and South Eastern Coastal Plain

Cuttack and Banki agricultural districts in Cuttack district, Jagatsinghpur, Kendrapara, Puri district, Khurda agricultural district in Khurda district and Berhampur, Chhatrapur agricultural districts in Ganjam district comprise this zone. The climate is hot and humid with mean rainfall 1489 mm and mean maximum as well as mean minimum temperature of 32.6^{0}C and 23.8^{0}C respectively. The soils of the zone are mainly loam, clay loam and coastal alluvial with a narrow strip of saline inundated areas. Rice, green gram, black gram, groundnut, jute, sesamum and vegetables are the major crops grown in this zone.

v. North Eastern Ghat

The zone consists of Khurda, Nayagarh, Kandhamal, Boudh, Gajapati districts, Rayagada and Gunupur agricultural districts in Rayagada district and Aska agricultural district in Ganjam district. The climate of the zone is hot, moist and sub-humid with mean annual rainfall of 1491 mm. The mean maximum and minimum temperature of this zone are 33.5^{0}C and 19.5^{0}C respectively. The soils of the zone are brown forest soil, sandy loam, loamy and clay with medium texture. Rice, ragi, black gram, green gram, til, groundnut, sugarcane, toria, turmeric and vegetables are the major crops grown in this zone.

vi. Eastern Ghat High Land

Koraput, Nawarangpur district except Dabugaon area are under this zone. The climate is warm and humid with mean annual rainfall of 1275 mm. The mean maximum and minimum temperature are 32.5^{0}C and 18.2^{0}C respectively. The zone consists mostly red soils with low organic matter content. Major crops like ragi, rice, small millets, vegetables, maize, arhar, niger, toria and sweet potato are grown in this zone.

vii. South Eastern Ghat

The agricultural district Jeypore in Koraput district and Malkangiri district are included in this zone. The zone is having warm and humid climate. The mean annual rainfall, maximum and minimum temperatures are 1614 mm, 34.0^{o}C and 17.6^{o}C respectively. Red and Red yellow mixed soils are the characteristics of this zone. Rice, arhar, niger, toria, maize, sweet potato and vegetables are the major crops grown.

viii. Western Undulating Zone

The zone comprises of the agricultural districts Bhawanipatna, Dharmagarh in Kalahandi district, Padmapur in Bolangir district, Khariar in Nawapara district and Dabugaon areas of Nawarangapur district. Hot, moist and sub-humid climate with mean annual rainfall of 1368.8 mm is the main features of the zone. The mean maximum and mean minimum temperature are 32.2^{o}C and 16.3^{o}C respectively. Black, heavy light soils and mixed red black soil with medium to heavy texture are found in this zone. Rice, ragi, arhar, bengal gram, toria, onion, cotton, vegetables and groundnut are the major crops grown in this zone.

ix. Western Central Table Land

Sambalpur, Baragarh, Bolangir and Sonepur districts belong to this zone. The climate is hot, moist and sub-humid with mean maximum temperature of 36.5^{0}C and mean minimum temperature of 15.8^{0}C. Red, heavy textured calcareous and light to medium textured soils are mostly seen in this zone. Average annual rainfall is 1357.2 mm. Major crops grown are rice, til, black ram, wheat, green gram, groundnut, sugarcane and vegetables.

x. Mid-Central Table Land

The zone consists of agricultural districts of Dhenkanal in Dhenkanal district, Angul and Kamakhyanagar in Angul district, Banki and Athagarh in Cuttack district and Sukinda block in Jajpur district. The climate of the zone is hot, dry and sub-humid. The mean annual rainfall is 1368 mm with mean maximum and mean minimum temperatures of 35.4^{0}C and 15.2^{0}C respectively. Light textured, lateritic, medium textured, red loam, mixed red and black soils are found in this zone. Rice, groundnut, sesamum, black gram and vegetables are the major crops grown in this zone.

3.3. Climate and environmental situation

The state lies in the sub-tropical belt. The chief characteristics of the climate are high temperature and high rainfall. Three defined seasons prevail in the state, i.e. Hot dry Summer – March to May, Monsoon – June to October and Winter season – November to February.

i. Rainfall

The average annual rainfall of the state is 1384.1 mm with an average of 69 rainy days, out of which around 1193 mm rain is received during monsoon season, i.e. June to October with average of 11 to 12 rainy days per month. The fluctuation in rainfall from year to year ranges from 600 to 800 mm in some districts. Drought, cyclone and flood are the usual calamities of the state. As the rainfall is concentrated during Kharif season, the concentration of cropping is therefore more during the season. The high co-efficient of variability in rainfall compels the farmers to bring the high lands under short duration rice or other low moisture requiring crops like millets, pulses, oilseeds, fiber crops or vegetables. The amount of rainfall during winter season is practically negligible.

ii. Temperature

Temperature is one of the limiting factors in crop production. The mean temperature, the range of diurnal variation, diurnal change and temperature summation influence both the duration and growth patterns of the crops. The mean annual maximum temperature varies from 27°C in January to 40°C in May. In case of northern plateau and other interior areas, even maximum temperature goes up to 44°C. The lowest temperature is recorded during later part of December and early part of January. The mean minimum temperature during the period varies from 10°C to 14°C in interior regions and 15°C to 19°C in coastal areas. The diurnal variation of temperature for most of the interior areas of the state is high and ranges from 15°C to 20°C except in monsoon when the value of the same varies from 10°C to 15°C.

iii. Humidity

The average annual relative humidity at 8.30 a.m. is 76.84%. Maximum humidity of 86.62% is recorded in July and minimum of 64.59% in May, the difference in annual range being only 12%. The central table land has the lowest humidity during March – May and is the driest part of the year. There is great variation in the humidity in different parts of the Ghat region. In Koraput district, the relative humidity ranges from 90% during monsoon to 68% during summer. The humidity in the coastal tract remains high throughout the year and never falls below 70% due to proximity of the sea.

The month wise average rainfall, number of rainy days, maximum and minimum temperature along with humidity is reflected in Table 3.

3.4. Evaporation

The rate of evaporation is an indicator of the vegetative possibilities of a given region. The rate of loss of water from plant surface exposed to the air is always lower than the rate of evaporation from an equal area of water surface under the same environmental condition. The total annual loss of water surface has been estimated to be 2142 mm or about 85 inches of water.

3.5 Sunshine

Light plays an important role in photosynthesis activities of plants. Performance of a crop variety is assessed by its capacity for minimum utilization of light energy. In Odisha, there are two distinct seasons of growing crops i.e. wet or kharif and dry or rabi season. Kharif season starts from June with nearly 13 hours of day length. Rabi season starts from January with 10 hours day length and extends towards summer season up to May with dry length of day around

Table 3: Monthwise climatic condition of Odisha

Sl.No.	Month	State Avg. Rainfall (mm)	Monthly Normal* rainfall (mm)	No. of normal rainy days	Mean Temp. (^{0}C)		Relative Humidity (%)	
					Max.	Min.	8.30 hrs	17.30 hrs
1.	January	52.7	11.4	0.7	27.06	14.50	81.74	70.37
2.	February	1.9	22.9	1.5	31.72	15.83	77.00	55.84
3.	March	0.6	25.5	1.7	36.00	18.05	69.50	47.89
4.	April	33.5	33.1	2.3	37.70	24.18	65.00	50.94
5.	May	21.5	63.3	3.8	39.94	26.44	64.59	47.12
6.	June	160.0	216.5	9.9	36.18	26.35	72.94	65.41
7.	July	326.7	339.9	15.4	31.13	24.63	86.62	77.88
8.	August	405.3	356.0	15.6	31.11	24.29	82.65	83.65
9.	September	232.5	231.9	11.5	31.53	24.41	84.88	73.52
10.	October	68.9	114.7	5.3	31.50	21.05	84.41	72.88
11.	November	77.3	31.5	1.3	28.88	17.65	79.70	70.11
12.	December	3.2	4.5	0.3	28.05	13.70	73.12	63.35
	Annual	1384.1	1451.2	69.3	32.57	20.92	76.84	64.91

* Average of last thirty years

Source: Odisha Agriculture Statistics,2012-13, Directorate of Agriculture, Orissa P. 1-7.

12 hours. The mean sunshine hours per day in Kharif ranges from 2.3 to 5.2 hours and in Rabi season 7.9 to 9.8 hours per day.

3.6. Soils of Odisha

The soils of different agro-climatic zones of Odisha were analysed under physiography. In general the residual soils of uplands are predominantly lateritic, highly leached, acidic (pH 5.0 ~ 6.50) and low in organic matter and nitrogen. The soil tends to be shallow, sandy and erodable in the uplands associated with low water holding capacity. The transplanted soils of the alluvial plains are moderately fertile, less acidic and generally more productive.

3.7. Water wealth

Odisha has enormous water wealth in its many rivers which are scattered all over the states. They cover an annual flow of 121 million ac-ft of water. Irrigation facilities have been expanded considerably since 1977. Over 33.66 lakh ha of land have been brought under irrigation through major, medium and minor irrigation projects including lift irrigation points. The crop wise irrigation potentialities in the state during the year 2012-2013 is presented in the following table which is self explanatory.

Though Odisha has a good source of water wealth, most of the interior districts are affected by drought regularly. The total cultivated area and irrigated area in different districts were also analysed and presented in table 5.

Table 4: Irrigated potentialities of the state (2012-2013)

Area in Lakh ha

Crops	Kharif			Rabi			Total		
	Total area covered	Area irrigated	% of Total area covered	Total area covered	Area irrigated	% of Total area covered	Total area covered	Area irrigated	% of Total area covered
Paddy	37.49	17.79	47	2.74	2.74	100	40.23	20.53	51
Other cereals	4.59	0.89	19	0.37	0.37	100	4.96	1.26	25
Pulses	7.23	0.40	6	13.19	1.77	13	20.42	2.17	11
Oilseeds	3.33	0.6	18	4.18	2.13	51	7.51	2.73	36
Fibers	1.48	0.07	5	-	-	-	1.48	0.07	5
Sugarcane	-	-	-	0.39	0.39	100	0.39	0.39	100
Vegetables	2.84	1.72	61	4.04	3.64	90	6.88	5.36	78
Spices	0.8	0.4	50	0.76	0.73	96	1.56	1.13	72
Others	-	-	-	0.02	0.02	100	0.02	0.02	100
Total	57.76	21.87	38	25.69	11.79	46	83.45	33.66	40

Source: Odisha Agriculture Statistics 2012-2013; p - v.

Table 5: Area under irrigation in different districts during 2012-2013

Sl. No.	District	Cultivated area (000'ha)		Irrigated area (000'ha)
		Kharif	Rabi	
1.	Balasore	250.0	125.65	91.23
2.	Bhadrakh	176.0	119.32	30.04
3.	Bolangir	346.0	56.88	37.82
4.	Subarnapur	128.0	76.14	39.74
5.	Cuttack	188.0	99.27	89.16
6.	Jagatsinghpur	104.0	58.74	34.54
7.	Jajpur	145.0	68.83	61.19
8.	Kendrapara	152.0	88.06	51.83
9.	Dhenkanal	186.0	76.20	30.10
10.	Angul	216.0	48.43	27.36
11.	Ganjam	406.0	160.75	48.11
12.	Gajapati	76.0	26.08	8.97
13.	Kalahandi	378.0	110.24	65.32
14.	Nawapada	189.0	46.11	15.25
15.	Keonjhar	298.0	79.47	34.76
16.	Koraput	304.0	87.95	58.42
17.	Malkangiri	142.0	61.18	22.37
18.	Nawarangpur	186.0	39.81	33.58
19.	Rayagada	193.0	40.45	20.73
20.	Mayurbhanj	437.0	119.64	53.45
21.	Kandhamal	128.0	26.07	10.32
22.	Boudh	89.0	28.76	12.08
23.	Puri	189.0	111.44	72.12
24.	Khurda	129.0	54.16	29.79
25.	Nayagarh	134.0	45.39	21.67
26.	Sambalpur	194.0	66.85	41.93
27.	Baragarh	349.0	151.08	93.66
28.	Deogarh	67.0	22.13	11.25
29.	Jharsuguda	88.0	14.21	7.92
30.	Sundergad	313.0	76.57	35.02
31.	Total	6180.0	2186.86	1178.73

Source: Odisha Agriculture Statistics 2012-2013 p-24-27.

3.8. Cropping intensity

Odisha agriculture statistics 2012-2013 revealed that out of 15,571 thousand hectares geographical area, 61.80 lakh hectares are cultivated lands, out of which 53.31 thousand hectares i.e. 86.26% are net sown area. Rice, pulses, oilseeds, jute, mesta, sugarcane, spices and vegetables are important crops grown. The high, medium and low lands of the state constitute about 47.15%, 28.40%, 24.45% respectively and about 50.6% of cultivated land is irrigated.

The cropping intensity of different districts of the state for 2000-2001, 2009-2010 and 2012-2013 is presented below in Table 6.

Table 6: Cropping intensity of various districts in Orissa

Sl.No.	Districts	Cropping intensity (%)		
		2000-2001	2009-2010	2012-2013
1.	Balasore	133	151	168
2.	Bhadrakh	124	132	135
3.	Bolangir	123	147	167
4.	Subarnapur	146	195	181
5.	Cuttack	147	195	193
6.	Jagatsinghpur	155	199	195
7.	Jajpur	167	195	195
8.	Kendrapara	159	184	191
9.	Dhenkanal	125	168	195
10.	Angul	132	165	168
11.	Ganjam	153	186	181
12.	Gajapati	154	196	197
13.	Kalahandi	139	169	183
14.	Nawapada	142	169	146
15.	Keonjhar	133	149	161
16.	Koraput	142	132	150
17.	Malkangiri	123	160	171
18.	Nawarangpur	136	155	160
19.	Rayagada	131	160	176
20.	Mayurbhanj	119	130	137
21.	Kandhamal	130	168	188
22.	Boudh	135	162	164
23.	Puri	168	210	187
24.	Khurda	143	179	192
25.	Nayagarh	146	204	198
26.	Sambalpur	121	168	154
27.	Baragarh	124	150	137
28.	Deogarh	112	189	177
29.	Jharsuguda	119	169	145
30.	Sundergarh	114	133	137
31.	Average	135	163	167

Source: Odisha Agricultural Statistics 2012-2013; p-18.

3.9. Land utilization pattern in Odisha

The land utilization pattern speaks about much of the agricultural extension planning in the state. Agriculture and land are not distinguishable separately. The total geographical area of the state remains the same 15571 thousand hectares. The development can be well judged from the type of land and its utilization for years together. The land utilization pattern for last 13 years has been depicted in table 7.

Table 7: Land utilization pattern in Orissa

(area in 000ha)

Sl. No.	Year	Geographical area	Forest area	Misc tree	Permanent pasture	Culturable waste	Land put to non-agril. Use	Barren & unculturable land	Current fallow	Other fallow	Net area sown
1.	2000-01	15571	5813	482	443	392	999	843	430	340	5829
2.	2001-02	15571	5813	482	443	392	999	843	320	434	5845
3.	2002-03	15571	5813	482	443	392	999	843	485	434	5680
4.	2003-04	15571	5813	482	443	392	999	843	369	434	5796
5.	2004-05	15571	5813	482	443	392	999	843	426	434	5739
6.	2005-06	15571	5813	482	443	392	999	843	474	434	5691
7.	2006-07	15571	5813	342	499	375	1298	840	526	229	5654
8.	2007-08	15571	5813	342	494	375	1298	840	556	229	5624
9.	2008-09	15571	5813	342	494	375	1298	840	576	229	5604
10.	2009-10	15571	5813	342	494	375	1298	840	606	229	5574
11.	2010-11	15571	5813	342	494	375	1298	840	773	229	5407
12.	2011-12	15571	5813	342	494	375	1298	840	888	229	5292
13.	2012-13	15571	5813	342	494	375	1298	840	849	229	5331

Source: Odisha Economic survey 2013-2014 Government of Orissa, p.111.

The previous table reveals that the area under permanent pasture, miscellaneous trees, cultivable waste are decreasing over years which can be inferred that land that were earlier being not in use are now used but may be due to lack of proper planning this land is not utilized for agriculture purpose. Rather there is a positive trend in forest area due to plantation programmes through several developmental schemes and departments. The tragedy is that much of the land is being put into non-agricultural use. Hopefully the agricultural use is not thought to be profitable and remunerative. It can also be said that due to lack of proper planning, valuation, vision and analysis of the future the land is not appropriately utilized and becomes barren and uncultivable and simultaneously remain fallow for years. The above situation as it reveals demand for a participatory and need based planning for the state.

3.10. Irrigation

In the absence of adequate irrigation facilities, agriculture is pathetically dependent upon the monsoons. As a result of the erratic behaviour of the monsoon, agricultural production fluctuates widely from year to year. The net irrigation potential created by the end of 2012-13 from all sources was 47.04 lakh hectares (31.03 lakh ha during Kharif and 15.74 lakh ha during Rabi) which is 50.6% of the total irrigable area of the state. Out of this, 1996.70 thousand hectares of land can be irrigated through major and medium irrigation projects, 671.60 thousand hectares through minor (flow), 860.78 thousand through minor (lift) and 1174.99 thousand through other sources which include private tanks, ponds, dug well, water harvesting structures and the like. However, the gross irrigated area during the year 2012-13 was 33.66 lakh ha (21.87 lakh ha during Kharif and 11.79 lakh ha during Rabi) which accounts to 71.5% of the irrigation potential created. This underscores the need for stepping after assured irrigation potential in order to isolate agriculture from the vagaries of monsoon.

Table 8: Net irrigation potential in the state through different types of sources

(Area in 000' hectares)

Sl. No.	Irrigation source	Irrigation potential created upto end of		
		2003-04	2009-10	2012-13
1.	Major and medium	1,234.86	1322.00	1996.70
2.	Minor (Flow)	497.49	551.11	671.60
3.	Minor (Lift)	383.80	492.94	860.78
4.	Other sources	572.58	596.16	1174.99
	Total	2,688.73	2962.21	4704.07

Source: Odisha Agriculture Statistics, 2012-13, Directorate of Agriculture & Food Production, Odisha, Bhubaneswar, p. 23

Table 8 mentioned above, clearly depicts a positive trend in net irrigation potential of the state that is a good sign for the agricultural scenario provided these resources are well planned and put to use most effectively. The crop and cropping pattern are to be well planned and need based as much as people oriented for better execution of the programmes for achieving desired end result.

3.11. Consumption of fertilizer

Fertilizer is one of the major production oriented inputs that decides the growth of agriculture in the state. As the table 9 shows there is an increase in the consumption rate but not appealing as in case of other states. The state ranks next to Kerala and Assam but is still much behind. May be it is due to low socio-economic status of the people in Odisha, but the cause is to be identified and definite planning is to be made for tracing out the reason and scope for increasing the fertilizer consumption rate that decides the yield of the major crops.

Table 9: Position of Odisha in Consumption of fertilizer at all India level

(Kg/hectare)

Sl.No.	Name of the state	Consumption of fertilizer				
		2008-09	2009-10	2010-11	2011-12	2012-13
1.	Andhra Pradesh	226.35	225.65	278.41	266.11	189.30
2.	Assam	57.47	63.11	67.59	67.25	66.26
3.	Bihar	171.56	165.60	180.58	184.24	212.23
4.	Gujarat	140.46	147.23	174.12	155.60	109.58
5.	Haryana	199.63	209.92	213.76	224.85	207.56
6.	Karnataka	142.08	159.55	163.94	181.43	117.23
7.	Kerala	94.50	95.94	106.23	112.90	104.71
8.	Madhya Pradesh	69.72	81.37	87.23	88.36	84.79
9.	Maharastra	113.27	135.32	152.81	133.65	102.73
10.	Odisha	62.00	59.78	62.85	62.25	58.74
11.	Punjab	224.59	237.08	242.73	243.56	250.19
12.	Tamil Nadu	217.58	205.82	220.60	227.01	164.58
14.	West Bengal	155.79	168.64	154.93	169.70	163.17
15.	Rajasthan	47.38	48.33	60.64	62.35	51.70
16.	Jharkhand	61.59	69.99	90.21	122.49	158.22
17.	Chhatisgarh	80.52	93.45	101.09	107.10	106.10
18.	Uttar Pradesh	161.78	170.96	174.28	169.91	183.23
	All India	127.21	135.27	146.32	139.67	128.34

Source: Odisha Agriculture Statistics, 2012-13 p.22.

Table 10: Status of operational holdings in the state

Sl.No.	Holding size	No. of Operational holdings			Area of Operation (in Hectares)		
		2000-01	2005-06	2010-11	2000-01	2005-06	2010-11
1.	Marginal(Below 1.0 ha)	2294520	2597164	3368296	1155145	1341668	1921842
		(56.42)	(59.62)	(72.17)	(22.73)	(26.73)	(39.53)
2.	Small(1.0 ha to 2.0 ha)	1113590	1156162	918647	1543709	1587713	1497752
		(27.38)	(26.54)	(19.68)	(30.38)	(31.63)	(30.80)
3.	Semi-medium(2.0 ha to 4.0 ha)	500531	472129	311261	1344201	1250650	918947
		(12.31)	(10.84)	(6.68)	(26.46)	(24.92)	(18.90)
4.	Medium(4.0 ha to 10.0 ha)	145110	119529	63688	817591	658208	381272
		(3.57)	(2.74)	(1.36)	(16.09)	(13.11)	(7.84)
5.	Large(10.0 ha and above)	13375	11408	5579	220387	181237	142019
		(0.33)	(0.26)	(0.11)	(4.34)	(3.61)	(2.93)
	Total	4067135(100.00)	4356392(100.00)	4667471(100.00)	5081033(100.00)	5019476(100.00)	4861832(100.00)

Figures in brackets indicate percentage to total
Source: Odisha Economic Survey 2013-2014, Government of Odisha p.125.

3.12. Operational holdings

The agricultural census 2010-11 shows that there are 46.67 lakh operational holdings in the state in comparison to 40.67 lakh as per 2000-01 census and 43.56 lakh as per 2005-06 census registering an increase of 14.75 % over a period of ten years. The total area of the operational holdings which was 50.81 lakh hectare in 2000-01 has marginally declined to 50.19 1akh hectares in 2005-06 and 48.62 lakh ha in 2010-11. The average size of holdings which was 1.25 hectare in 2000-01 has decreased to 1.15 hectares in 2005-06 and 1.04 ha in 2010-11.

3.13. Livestock population

Development of animal resource has considerable relevance and significance for development of an agrarian economy. Next to crop production, animal husbandry is the most important income generating activity in rural households. The distribution of livestock population in the state as per Livestock census 2007 is given in table 11.

Table 11: Livestock population in Odisha

(in lakh nos.)

CensusYear	Cattle		Buffaloes	Goats	Sheep	Pigs	Total
	Total	Crossbred					
2003	142.81	10.43	14.39	59.74	17.59	5.69	240.22
2007	123.09	17.03	11.90	71.27	18.18	6.12	230.57

Source: Economic Survey 2013-2014, Government of Orissa p. 133.

It is observed that the percentage of cross breed/ improved animals in Odisha state is very poor, that is out of total cattle population the percentage of cross bred cattle is 13.83% and in case of buffalo the percentage of improved buffalo is 6.8% only.

3.14. Availability of animal products

The per capita availability of animal products has by and large gone up steadily. The per capita availability of milk is 117g/head/day while that of egg is 55 eggs/ head/ annum during 2012-13. The production of milk and egg was 1.651 million tonnes and 2319 million numbers in our state respectively during 2009-10. However, the state agricultural policy 2014 aims at increasing the production and productivity in dairy, poultry and encourages other animal husbandry activities through scientific breeding, balanced feeding, adequate health care, better marketing facilities and improved management practices.

Table 12: Year wise production and per capita availability of milk, meat and eggs in Odisha

Year	Production of milk (Thousand MT)	Per capita availability of milk (g/day)	Production of meat (Thousand MT)	Per capita availability of meat(kg/annum)	Production of eggs (Millions)	Per capita availability of eggs (Nos./annum)
2005-06	1342.30	98	52.04	1.385	1278.71	34
2006-07	1424.64	103	55.01	1.450	1424.64	38
2007-08	1620.44	115	58.82	1.530	1549.48	39
2008-09	1598.05	113	115*	2.933	1993.26**	51
2009-10	1651.33	112	130*	3.217	2319.32**	57
2010-11	1670.00	109	138	3.290	2357.10	56
2011-12	1718.00	112	138	3.29	2300.70	55
2012-13	1784.00	117	141.83	3.38	2322.90	55

* include poultry meat

** include private sector production

Source: Odisha Economic Survey 2013-2014, Government of Odisha p.136.

3.15. Fisheries resources

Fishery, an allied sector of agriculture is gradually drawing attention as an important area for employment and income generation. Odisha being abundant with water resources has enough scope and potentialities for development of fisheries sector. The state is one of the important maritime states of India having excellent scope of potentialities for development of fisheries sector. It has a coastline of 480 kms with continental shelf area of 24,000sq. kms along the Bay of Bengal. It is also endowed with brackish water area of 4.18 lakh hectares and fresh water area of 6.75 lakh hectares. Odisha ranks 9th in total production of fish amounting to 386.19 thousand MT contributing to 4.71% of total production of the country (8230.71 thousand MT) during 2010-11. The fish production in the state during 2012-13 was 410.14 TMT of which 291.83 TMT came from inland sources and 118.31 TMT was from marine sources with a value of 3009.74 crores. The value of fish production thus has increased by 12.49% over 2010-11 (2748.44 crore rupees).

3.16. Forest resources

Forests play a vital role in maintaining ecological stability. They provide protection against and reduce impact of natural calamities like droughts, floods and cyclones. Forests are important source of livelihood for weaker section of society in general and tribals in particular. Forests also provide a wide variety of products of commercial value such as timber lac, resins, oilseeds, kendu leaves and medicinal plants.

Odisha has a record of forest area measuring 58,135.47 sq. km including 26,329.12 sq. km of reserve forest, 11,722.89 sq. km of demarcated and protected forests, 3838.78 sq. km of un -demarcated protected forest, 20.55 sq. km of un-classified forest and 16261.34 sq km of other forests including village and private forests which is 37.34% of the state's geographical area of 1,55,707 sq. km during 2011-12. As per the 'State of Forest Report - 2009' prepared by the 'Forest Survey of India', the state has 48,855 sq km of forest cover including 219 sq km of mangrove forests and 5782 sq km of shrub.

Forests of the state are unevenly distributed. The coastal districts with higher population density are having smaller areas of forest. Higher forest coverage is found in the district of Kandhamala (71.19%) followed by Malkangiri (57.95%), Gajapati (57.09%), Deogarh (53.07%) and Nayagarh (53.05%) during 2011-12. The National Forest Policy 1988 has mandated that 33% of the geographical area should be under forest cover, with 60% coverage in hilly tracks and 20% in plains. Against this the state has 37.33% forest cover according to

Table 13: Total Production of fish in Odisha

(in Thousand MT)

SI. No.	Year	Inland fish production			Marine fish production	Total Production	Per capita fish consumption (kg)
		Freshwater	Brackishwater	Total			
1.	2007-08	195.70	23.00	218.70	130.80	349.50	9.29
2.	2008-09	213.00	26.30	239.30	135.50	374.80	13.27
3.	2009-10	215.80	25.50	241.30	129.30	370.60	13.27
4.	2010-11	224.96	27.75	252.71	133.48	386.19	9.42
5.	2011-12	237.47	30.06	267.33	114.30	381.83	10.14
6.	2012-13	261.92	29.91	291.83	118.31	410.14	9.13

Source: Odisha Economic Survey 2013-14, Government of Odisha p.100,143.

Forest Survey of India. However, still there is an imperative need to accelerate efforts of afforestation and regeneration of the degraded forests to improve upon the climatic condition and natural environment as well.

4

Research Methodology

This chapter "Research Methodology" deals with methods and procedures followed in conducting the study. The chapter contains three parts. The first part deals with the sampling technique and the second part covers the empirical measures of different variables and last part describes details of data collection and statistical measures used in the study.

4.1. Sampling technique

4.1.1. Selection of districts

The study is designed to investigate the sustainability of small farms in coastal eco system of Odisha. The major concern of sustainability is observed in coastal tract like, the problem of salinity, flood, depletion of ground water etc are observed. Any attempt to work towards sustainability in agriculture has to look into land status of the state. Odisha has land problems in respect to salinity, low land, water merged and other different types resulting in low level of productivity.

Table 14: District wise distribution of coastal saline lands in Odisha with present land use

SI.No.	District	Land use (Area in ha.)			
		Agriculture	Mangrove	Idle	Total
1.	Ganjam	6,480	—	3,465	9,94
2.	Puri	43,930	3,250	13,076	60,256
3.	Cuttack	67,342	18,667	9,440	95,449
4.	Balasore	75,679	3,249	10,543	89,471
	Total	1,93,431	25,166	36,524	2,55,121

Again water logging is a problem in coastal districts. The waterlogged area is found to be highest in the district of Cuttack followed by Puri, Balasore and Ganjam.

Keeping these considerations in view, after careful examination of situations, four districts namely Puri, Khurda, Ganjam and Balsore of the state were selected. Districts were selected considering seriousness of land problem, water logging, salinity and frequent occurrence of natural calamities.

4.1.2. Selection of blocks

Selection of block was done purposively. From each district, one block was selected to carry out the study. The blocks covered under study were, Pipili of Puri district, Balipatna of Khurda district, Hinjilicut of Ganjam district and Basta of Balasore district.

4.1.3. Selection of villages

Two representative villages from each block were selected. The selection of villages was based on considerations like, (a) magnitude of problem, (b) co-operation of people, (c) scope for improvement in ecology, (d) medium productivity level of crops and (e) no such studies carried out earlier in these villages.

Table 15: Analysis of water logging in the coastal districts

Area in '000 ha

SI.No.	Name of the district	Net Area Sown	Irrigated area	% of area under irrigation in composition to net area sown	Total land water logged	% of Water logged area in relation to irrigated area
1.	Cuttack	705	242	34.32	21.36	8.82
2.	Puri	466	169	36.26	13.95	8.25
3.	Balasore	424	71	16.74	6.36	8.95
4.	Ganjam	501	226	45.10	5.86	2.58

Source: Census of India 2011, Odisha Series

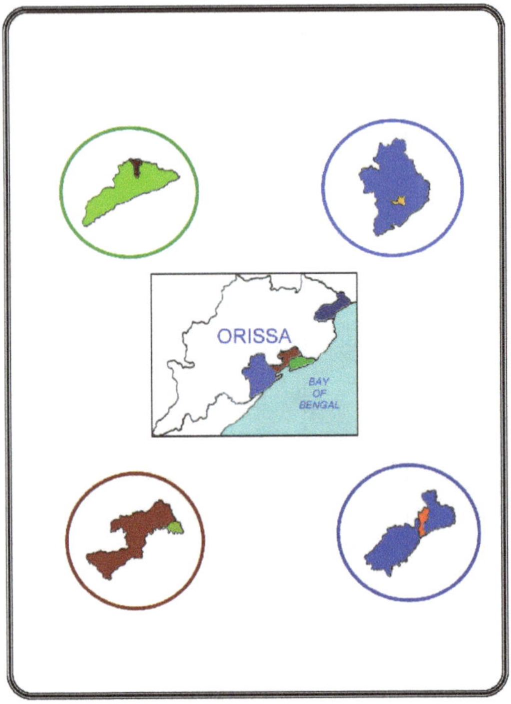

Fig. 1: Map showing area of the study

Table 16: Villages, Blocks and Districts under study

District	Block	Village
Puri	Pipili	Aruha Bantalsingh
Khurda	Balipatna	Alasi Rajasa
Ganjam	Hinjilicut	Sampur Makarjhola
Balasore	Basta	Sadanandapur Baunsamukha

4.1.4. Selection of sample respondents

The study considered only small and marginal farm operators as per definition of State Govt. of Odisha. From the list available at block level, a total of 200 respondents at the rate of 25 from each village (Table 16) were finally selected for interview and data collection. The study encompasses 200 respondents selected randomly from each village covering 8 villages of 4 blocks of 4 districts.The basic facts of selected 8 villages under study were given in Table 17.

Table 17: Basic facts about the villages under investigation

Sl. No.	Facts	Pipili Block		Balipatna Block		Hinjili Block		Basta Block	
		Aruha	Bantalsingh	Alasi	Rajasa	Sampur	Makarajhola	Sadanandapur	Baunsamukha
1.	Geographical area (ha)	143.2	195.0	126.0	498.4	216.0	516.0	184.6	87.2
2.	Cultivable area (ha)	130.0	183.4	106.0	467.0	210.0	464.0	170.0	69.0
3.	Cultivated land (ha)	124.0	180.0	106.0	467.0	210.0	464.0	170.0	69.0
4.	High land (ha)	28.0	30.0	9.0	10.0	43.0	72.0	10.0	10.0
5.	Medium land (ha)	80.0	80.0	34.0	200.0	95.0	145.0	98.0	21.0
6.	Low land (ha)	16.0	70.0	63.0	257.0	72.0	246.0	62.0	38.0
7.	Irrigated land (ha)								
	(i) Kharif	103.0	168.0	85.0	102.0	189.0	423.0	100.0	40.0
	(ii) Rabi	70.0	47.0	42.0	60.0	20.0	40.0	78.0	21.0
8.	Total no. of families	186	251	83	156	297	726	222	261
9.	Total no. of farm families	158	216	58	119	246	534	187	196
10.	Marginal farm families	77	116	28	56	150	258	100	142
11.	Small farm families	45	48	22	38	51	250	55	32
12.	Medium farm families	21	34	—	—	—	—	22	6
13.	Large farm families	15	18	8	25	45	26	10	16
14.	Landless Labour	28	35	25	36	51	192	35	65
15.	Respondents selected	25	25	25	25	25	25	25	25
16.	Percentage sample to farm families	6.32	8.64	43.10	21.00	10.16	4.68	13.37	12.80

Source: Census of India 2011, Odisha Series

4.2. Selection of variables

Independent variables included in the study were selected on the basis of review of literatures, discussion with the experts, judges rating method and preliminary study conducted in the area of investigation. Only those variables, which were found having relevance with sustainability of agriculture were included within the framework of the study.

The variables grouped under independent factors and their relationship with dependent variables on sustainability of agriculture were invariably studied in all the aspects. As per interview schedule, the variables have been arranged under six parts.

Part-I : Socio-personal information (Personal, social, economic, communication, marketing, farming behaviour and training)

Part-II : Awareness on organic farming

Part-III : (A) Factors of sustainability

(B) Perception about sustainability

(C) Sub-factors of sustainability

Part-IV : Adoption of recommended sustainability measures

Part-V : Agribusiness opportunities for sustainable agriculture

Part-VI : (A) Extension and communication elements

(B) Suggestions.

The scoring of different variable was made as described below. The schedule meant for experts consisted of two parts. The first part was dealt with factors of sustainability of agriculture and the second part was used to obtain suggestions to attain sustainability of agriculture.

4.3. Empirical measurements of variables

Variables used in the study were quantified depending on their relative merit in qualifying the measurements keeping objectives of the study in view and scoring method was adopted to measure the variables.

Table 18: Variables and quantification

Sl.No.	Variable	Groups under variables	Score assigned
1.	Age	(i) up to 30 years	1
		(ii) 31-40 years	2
		(iii) 41-550 years	3
		(iv) Above 50 years	4
2.	Education	(i) Illiterate	0
		(ii) Primary School	1
		(iii) M.E.School	2
		(iv) High School	3
		(v) College	4
3.	Family size	(i) within4 members	1
		(ii) More than 4 members	2
4.	Family type	(i) Nebulous family	1
		(ii) Joint family	2
5.	Caste	(i).ST	1
		(ii) Sc	2
		(iii)OBC	3
		(iv) Others	4
6.	Land	(i). Up to 1 acre	1
		(ii). 1.1 to 2 acre	2
		(iii). 2.1 to 3 acre	3
		(iv). 3.1 to 4 acre	4
		(v). 4.1 to 5 acre	5
7.	Occupation	All respondents have farming as major occupation	1
8.	Irrigation	(i). Yes	1
		(ii).No	2
9.	Crops grown	(i).Mono crop	1
		(ii).Two crops	2
		(iii). Three crops	4
10.	Livestock	(i) up to three	1
		(ii) More than three	2
11.	Farm energy use	(i) Yes	2
		(ii) NO	I
12.	Farm implements	(i) up to 4	1
		(ii) More than 4	2
13.	Communication Behavior	(i) Exposed to media for farm information	1
		(ii) Not exposed	0
14.	Place of marketing	(i) Own locality	1
		(ii) Closed to locality	2
		(iii) Outside locality	3
15.	Opinion about sale price of produce	(i) Remunerative	4
		(ii) Remunerative at times	3
		(iii) Not remunerative	2
		(iv) Not concerned	1

Sl.No.	Variable	Groups under variables	Score assigned
16.	Farm produce	(i) Just manageable	1
		(ii) Little over family maintenance	2
		(iii) More than family requirement	3
		(iv) Only grow for market	4
17.	Experience in farming	(i) Up to 10 years	1
		(ii) 11-20 years	2
		(iii) More than 20 years	3
18.	Training	(i) Yes	1
		(ii) No	0

4.4. Perception about sustainability in farming

Variables relating to perception of the respondents about sustainability in farming used in the present study were quantified following the method of Liker T test through a 5-point scale.

Table 19: Perception about sustainability in farming

Variables	Strongly agree	Agree	Undecided	Disagree	Strongly disagree
1.Ecological soundness	5	4	3	2	1
2.Economic viability	5	4	3	2	1
3.Social accessibility	5	4	3	2	1
4.Humanness	5	4	3	2	1
5. Adaptability	5	4	3	2	1

4.5. Sustainable agriculture index (SAI)

SAI has been worked out on the basis of 5 important parameters i.e. ecological soundness, economic viability, social accessibility, humaneness and adaptability and taking their relative position with assigned score and making it average.

$$SAI = \frac{\text{Scores obtained X 100}}{\text{Total No. of statements with Maximum Obtainable Score}}$$

4.6. Organic farming

It was studied to find out the awareness, knowledge and benefits of organic farming towards sustainability of small farms. Problems and adoption of organic farming and suggestions there to were obtained. They were ranked through assigning scores 3, 2 and 1 to very important, important and less important respectively.

4.7. Adoption of recommended measures of sustainable agriculture

This has been studied in reference to whether farmers adopt or do not adopt the recommended measures of sustainable agriculture. Score assigned is 1 for adopted and 0 for not adopted.

4.8. Agribusiness opportunities

Various agribusiness opportunities were sorted out and grouped under human factors, physical aspects, animal husbandry, farm machineries, and financial aspects related to sustainable agriculture. In this case scores were assigned in a 3 point scale is 3, 2 and 1 for full, partial and not at all. The gap % was also analysed.

4.9. Ranking of sub-factors of sustainability of agriculture

The sub-factors of sustainability of agriculture were ranked through assigning scores 3, 2 and 1 to very important, important and less important respectively.

4.10. Extension approaches

The suggestive extension measures to enhance sustainability in agriculture were studied under components of social structure, cooperation mechanism, evaluation technique and marketing network. The sub-components under each component was highlighted in order of importance through ranking.

4.11. Suggestions

Suggestions of farmers as well as experts were recorded under the components of ecological soundness, economic viability, social accessibility, humaneness and adaptability.

4.12. Statistical methods used

The statistical methods such as percentage, mean score, frequency, rank order analysis, Pearson coefficient of correlation, student t-test, Liker-T test, gap analysis, analysis of variance, Duncan's multiple range test, critical ratio test, χ^2 test, multiple correlation, factorial r matrix, Spearman rank order correlation (rho), Multiple correlation were used in the study for analysis of responses based on the nature of the data and type of information required.

5

Farmers of Small Farm Production System

Social research concentrates more on socio economic variables to find out effects of the events that take place in social system. The magnitude of variation in the variables assumes to bring variation on the effect. The variables relating to farmers of small farms are many and they are specific to farming system of the coastal ecosystem. Such variables were grouped under personal, social, economic, communication behaviour, marketing behaviour, farming behaviour and training. The sub variables under each heading were given in the following table.

Table 20: Profile of sample under study

Sl.No.	Major variables	Sub-variables
1	Personal	1. Age 2. Education
2.	Social	1. Family composition 2. Family type 3. Caste 4. Membership in formal organization
3.	Economic	1. Land holding 2. Occupation 3. Irrigation 4. Crops grown

(Contd.)

Sl.No.	Major variables	Sub-variables
		5. Livestock 6. Additional source of income 7. Use of energy 8. Farm implements
4.	Communication behaviour	1. Newspaper 2. Radio 3. Television
5.	Marketing behaviour	1. Marketable produce 2. Opinion about sale price 3. Place of marketing
6.	Farming behaviour	1. Experience in farming 2. Farming responsibilities
7.	Training	1. Training received 2. Use of training knowledge

Summarizing the major factors and sub-factors, the total independent variables under study were as follows:

Table 21: Independent variables of sample profile

Sl. No.	Independent variables	Number of Sub-variables
1.	Personal	2
2.	Social	4
3.	Economic	8
4.	Communication behaviour	3
5.	Marketing behaviour	3
6.	Farming behaviour	2
7.	Training	2
	Total	24

Further, as per the requirement of the study the profile of the sample was examined in connection with the different perception levels about the sustainability of agriculture. It is so because all the variables identified mainly, personal, social, economic, communication behaviour, marketing behaviour, farming behaviour and training have direct bearing on the formation of perception about any event particularly sustainability in agriculture in which they reside and work.

5.1. Level of perception

The level of perception was decided taking difference between maximum and minimum score obtained by individual sample. As much as 21% of the differential score constituted the low perception of sample, where as more than 25% high level and leaving the rest in between i.e., under medium level. The analysis so obtained is classified in the following table.

Table 22: Classification of sample based on level of perception (n=200)

Level of perception	Score range	Frequency	Percentage
Low	Up to 94	42	21
Medium	95 to 106	108	54
High	107 and more	50	25

The classification of sample on the basis of score obtained against core items of sustainability like ecological soundness, economic viability, social accessibility, humaneness and adaptability were taken into considerations assigning score on a five point scale. The relationship between profile of sample and level of perception has been explored.

5.2. Personal factors

i. Age group

Many of the extension studies reported that with increase of the age the knowledge and adoption level increases resulting in perceptiveness of the individual. While examining the hypothesis, the study made classification of sample into four different groups and explored the relationship with the level of perception about sustainability of agriculture.

High level of perception is found with the age group of 41 to 50 years while the respondents below 30 years and above 51 years have low and medium level of perception. The Chi- square value being significant confirms that sample of higher age group have better understanding of sustainability owing to their experience, working in agriculture and exposure to farming world.

ii. Educational status

Sociological studies try to explore the impact of education on various aspects of human development. Education is one of the most important factors to bring people to the world of progressiveness. The present study attempted to ascertain influence of educational status on perception level about sustainability. The education of respondents in relation to positive perception about sustainability reveals a significant relationship.

Highest percentages of respondents (37.5%) are educated up to M.E. School level followed by primary level to an extent of 29.5%. The respondents at high school level are observed to be 18% followed by 9.5% at college level. However,

Table 23: Age and perception on sustainability about agriculture (n=200)

Sl. No	Age group (years)	Low		Medium		High		Total		χ^2 Value
		F	%	*f*	%	*f*	%	*f*	%	
1.	Up to 30	0	0	0	0	0	0	0	0	13.505 *
2.	31 to 40	1	2.44	4	3.70	2	4.00	7	3.50	
3.	41 to 50	17	40.50	63	58.33	38	76.00	118	59.00	
4.	51 and above	24	57.14	41	37.97	10	20.00	75	37.50	
5.	Total	42	100.00	108	100.00	50	100.00	200	100.00	

* Significant at 5% level of probability

Table 24: Education with perception on sustainability of agriculture (n=200)

Sl. No	Education	Low		Medium		High		Total		χ^2 Value
		F	%	*f*	%	*f*	%	*f*	%	
1.	Illiterate	9	21.43	1	0.93	1	2.00	11	5.50	43.32*
2.	Primary	8	19.05	42	38.89	9	18.0	59	29.50	
3.	M.E. School	18	42.86	39	36.11	18	36.0	75	37.50	
4.	High school	7	16.67	17	15.74	12	24.0	36	18.00	
5.	College	0	0	9	8.33	10	20.0	19	9.50	
6.	Total	42	100	108	100	50	100	200	100	

* Significant at 5% level of probability

the percentage of illiterates was minimum (5.5%). Comparing three groups of respondents in relation to perception level, higher trend in position was observed with rise of educational level. The χ^2 test being significant confirmed that perception is closely associated with level of education.

5.3. Social factors

The social factors included in the study consisted of family composition, type of family, and membership in formal organization.

i. Family composition

The concept of small family has been accepted by majority of our population permitting reduction in family size. According to sociological studies composition of family and cohesiveness in interaction have direct influence on progressiveness and attitude of individuals leading to positive perception. While examining family composition of sample under study, in group of up to 4 and above 5 members per family, the analysis reveals that 86.5% of the samples have the family with 5 members on an average while 13.5% that have up to 4 members per family. Similar trend in case of three levels of perception was revealed. The χ^2 test being non-significant leads to infer that family composition is independent of perception level about sustainability in agriculture.

ii. Family type

Family composition and family type are interrelated in many of the social aspects. The present day living tends to be nucleus type over the joint family system. The joint system of living is declining at present. Table 26 contains information about type of family of the sample under study.

The results contained in table 26 presents a clear picture of family type under low, medium and high level perception groups. The joint family system is higher (55.56%) over nucleus one in case of medium group. On the whole, the nucleus family composition is found to be higher (53.0%) over joint system one (47.0%). The χ^2 value being significant reveals a positive association between type of family and level of perception of the farmers in the matter of sustainability of agriculture.

iii. Caste

The caste continued to be one of the influential factors of activities of individuals in the society. Many studies have proved beyond doubt that influence of caste on individual in society continues to play important role. The present study while attempting to analyze relationship between level of perception and caste structure the following results were obtained.

Table 25:Composition of family and perception of sustainability in agriculture (n=200)

Sl. No	Family composition	Low		Medium		High		Total		χ^2 Value
		F	%	*f*	%	*f*	%	*f*	%	
1.	Up to 4	8	19.05	12	11.11	7	14.0	27	13.50	1.645N.S.
2.	More than 5	34	80.95	96	88.89	43	86.0	173	86.50	
3.	Total	42	100	108	100	50	100	200	100	

N.S. Non significant

Table 26: Family type and perception of sustainability in agriculture (n=200)

Sl. No	Family type	Low		Medium		High		Total		χ^2 Value
		F	%	*f*	%	*f*	%	*f*	%	
1.	Nucleus	26	61.90	48	44.44	32	64.0	106	53.0	191.45*
2.	Joint	16	38.10	60	55.56	18	36.0	94	47.0	
3.	Total	42	100	108	100	50	100	200	100	

* Significant at 5% level of probability

Table 27: Caste and perception of sustainability in agriculture (n=200)

Sl.No	Caste	Low		Medium		High		Total		χ^2 Value
		F	%	*f*	%	*f*	%	*f*	%	
1.	Scheduled tribe (ST)	0	0	0	0	0	0	0	0	23.98*
2.	scheduled caste (SCT)	14	33.33	8	7.41	2	4.00	24	12.00	
3.	Other back word classes (OBC)	14	33.33	48	44.44	20	40.00	82	41.00	
4.	Others	14	33.34	52	48.15	28	56.00	94	47.00	
5.	Total	42	100	108	100	50	100	200	100	

* Significant at 5% level of probability

The sample consisted mostly of OBC (other back word classes) and other caste to the tune of 41% and 47% respectively with SC (scheduled caste) up to 12%. Further, it is observed that OBC and other caste dominated under medium level of perception. The Chi square test being significant (χ^2=23.98), the inference is that the caste structure influences the formation of perception of the small farm operators on sustainable in agriculture.

iv. Membership in formal organization

Today in rural Odisha, innumerable organizations are coming up to participate in development process. The small farmers are to become members to fulfill the professional as well as social requirements. On the other hand, membership in organization is an indication of progressiveness. Examining the situation with the sample under study it is revealed that as much as 47% of sample are members in three to four organizations followed by 33% up to two organizations while 20% did not have any membership. Moreover, all the respondents under high level of perception were members where as non-members were found in case of low and medium perception level. The Chi square value being significant ($\div^2$= 44.186), membership is found to be related to perception level of the sample farmers.

5.4. Economic factors

The study surveyed the economic variables of the small farm operators. So far as economic factors are concerned, considering the relevance of the sample, eight economic factors consisting of land holding, occupation, irrigation, crops grown, livestock, additional source of income, use of energy and farm implements were taken into consideration.

i. Land holding

The households of farming communities are classified into small, marginal and large farm families by the Government of Odisha on the basis of their land holdings. The families possessing one to two hectors of land are further classified at micro level. The study attempted to examine distribution of sample on five levels of land holding starting from one acre to 5 acres of land.

The distribution of sample reveals that maximum respondents have land within 3.1 – 4.0 acres followed by 2.1 to 3 acres and 1.1 to 2.0 acres. However, the data under Chi square test is found to be non-significant. The homogeneity of sample on landholding size irrespective of districts is the reason accounted for this. It is so because the sample was drawn as per classification of State Government on the specifications of parameters of small and marginal farm families.

Table 28:Membership in formal organization and perception of sustainability in agriculture (n=200)

Sl. No	Membership	Low		Medium		High		Total		χ^2 Value
		F	%	*f*	%	*f*	%	*f*	%	
1.	Up to 2	12	28.57	33	30.56	21	42.0	66	33.00	24.186*
2.	3 to 4	13	30.95	52	48.15	29	58.0	94	47.00	
3.	No membership	17	40.48	23	21.29	0	0	40	20.00	
4.	Total	42	100	108	100	50	100	200	100	

* Significant at 5% level of probability

Table 29: Land holding and perception of sustainability in agriculture (n=200)

Sl. No	Land holding (Ac)	Low		Medium		High		Total		χ^2 Value
		F	%	*f*	%	*f*	%	*f*	%	
1.	Up to 1.0	4	9.52	1	0.93	0	0	5	2.50	14.3587N.S.
2.	1.1 ~ 2.0	15	35.71	32	29.63	14	28.0	61	30.50	
3.	2.1 ~ 3.0	12	28.57	35	32.41	15	30.0	62	31.00	
4.	3.1 ~ 4.0	9	21.43	24	22.22	13	26.0	46	23.00	
5.	4.1 ~ 5.0	2	4.76	16	14.81	8	16.0	26	13.00	
6.	Total	42	100	108	100	50	100	200	100	

N.S-Non Significant

ii. Occupation

Occupation of small farmers is farming. With emphasis on diversification, there has been difference at micro level in the matter of occupational profession. It is of common observation at villages that families earn their livings from more than one occupations like business and service. In finding out the distribution of sample under service, business and farming occupations the results show that occupation as a variable is found to be independent of perceptional level about sustainable agriculture. The distribution of sample against the occupations of service, business and farming was found to be 8.50%, 4.50% and 87.0% respectively. Invariably, the distribution of sample is found to be higher in case of farming irrespective of level of perception.

iii. Area under irrigation

It was observed during investigation that small farmers in coastal tract of Odisha have some land under irrigation facilities either for crops or vegetables using various sources. The sample being homogenous in nature and irrigation found to be constant factor with the entire sample (100%). Any differential response was not obtained. This leads to conclude that all small farmers under study have area under irrigation irrespective of magnitude and crops.

iv. Crops grown

The cropping intensity in the state is 148% which indicates that more than one crop is taken in many areas. The coastal ecosystem of state is enriched with availability of underground water. Under ground water sources tend farmers to take more than one crop per year. Analysis of data collected in this context appears in the Table 31.

Findings reveal that majority of sample (75%) of small farm operators take two crops per year followed by 14.5% who practice mono crop and 10.5% take more than three crops. The χ^2 test was found to be non-significant, owing to homogeneity in cropping pattern. In coastal area of the state under study, it is inferred that taking number of crops per year and formation of perception are not positively related.

v. Animal wealth

Livestock is an integral part of farming system in the state of Odisha. It is normally observed that those who undertake farming must have livestock of any kind to support farm operations. While examining the situation of sample under study, the results obtained are presented in Table 32.

Table 30: Occupation and perception of sustainability in agriculture (n=200)

Sl. No	Occupation	Low		Medium		High		Total		χ^2 Value
		F	%	*f*	%	*f*	%	*f*	%	
1.	Service	3	7.14	8	7.41	6	12.00	17	8.50	1.787 N.S.
2.	Business	1	2.38	6	5.56	2	4.00	9	4.50	
3.	Farming	38	90.48	94	87.03	42	84.00	174	87.00	
4.	Total	42	100	108	100	50	100	200	100	

Table 31:Crops grown and perception of sustainability in agriculture (n=200)

Sl. No	Crops grown	Low		Medium		High		Total		χ^2 Value
		F	%	*f*	%	*f*	%	*f*	%	
1.	Mono crop	4	9.52	16	14.81	9	18.00	29	14.50	3.026N.S.
2.	2 crops	35	83.33	78	72.22	37	74.00	150	75.00	
3.	3 crops	3	7.14	14	12.96	4	8.00	21	10.50	
4.	Total	42	100	108	100	50	100	200	100	

N.S-Non Significant

Table 32: Animal wealth and perception of sustainability in agriculture (n=200)

Sl. No	Animal wealth	Low		Medium		High		Total		χ^2 Value
		F	%	*f*	%	*f*	%	*f*	%	
1.	Less than 4 animals	12	28.57	32	29.63	17	34.00	61	30.50	0.396N.S.
2.	4 & above 4 animals	30	71.43	76	70.37	33	66.00	139	69.50	
3.	Total	42	100	108	100	50	100	200	100	

N.S-Non Significant

As much as 69% possess more than 4 animals against 34.5% who have up to 4 animals. The possession is higher in case of low perception group followed by medium and high. The occupation relating livestock does not get affected so much as in case of crops may be the reason for distribution of high percentage of respondents in low and medium perception group. In other words, respondents coming under low perception group keep more animals that support their farming operations. However, χ^2-test is not significant revealing any relationship between animal wealth and perception about sustainability in agriculture.

vi. Additional source of income

Additional source of income is related to diversifications in the traditional methods of farming and occupation. With increase of extension inputs there has been a change in traditional system of farming adding livestock and fishery as source of additional income. Moreover, livestock and fishery are important components of farming system in coastal tract of the state as revealed .

Of the sample 87.5% have only livestock to support farming and income where as 12.5% have fishery in addition to livestock. It is more in case of respondents of medium and high level of perception. The χ^2 value is not significant owing to reasons of limited farmers have fishery as additional component of family income.

vii. Use of energy

Use of energy is of recent importance in farming system of the state. With introduction of new agriculture policy and increase in subsidy for the farming sector energy use in form of electricity, diesel or petrol has been increased over the past years. The study attempted to find out use of energy among the small farm operators in coastal ecosystem. It is observed that out of sample, only 24.5% use energy against 75.5% who do not use energy in farm operations.

The results show that there is significant relationship between use of energy and perception about sustainability in agriculture. However, the per capita consumption of energy has not taken into considerations which would have been revealed additional information.

viii. Farm machinery

Use of farm machinery is given more importance to meet the shortage of labor in the state and reduce the cost of production. With increasing emphasis on agro-industries in the state, the use of tractor, power tiller, winnower, reaper and sprinkler etc. has been increased invariably. Moreover, there is also gap in use and possession of farm machinery as many prefer to hire these at the time of need. The concept has been further enriched by establishment of agro-clinic

Table 33: Additional source of income and perception of sustainability in agriculture

(n=200)

Sl.No	AdditionalIncome source	Low		Medium		High		Total		χ^2 Value
		F	%	*f*	%	*f*	%	*f*	%	
1.	Only livestock	40	95.24	94	87.04	41	82.00	175	87.50	3.703N.S.
2.	Livestock & fishery	2	4.76	14	12.96	9	18.00	25	12.50	
3.	Total	42	100	108	100	50	100	200	100	

N.S-Non Significant

Table 34: Use of energy and perception of sustainability in agriculture (n=200)

Sl. No	Energy use	Low		Medium		High		Total		χ^2 Value
		F	%	*f*	%	*f*	%	*f*	%	
1.	Yes	2	4.76	31	28.70	16	32.00	49	24.50	11.398*
2.	No	40	95.24	77	71.30	34	68.00	151	75.50	
3.	Total	42	100	108	100	50	100	200	100	

* Significant at 5% level of probability

at different parts of the state. The study attempted to find out possession of farm machinery among the sample as contained in the table below.

Table 35: Possession of farm machinery in agriculture (n=200)

Sl. No	Name of farm machinery	Frequency (*f*)	Percentage (%)
1.	Tractor	6	3.00
2.	Power tiller	18	9.00
3.	Reaper	0	0
4.	Thresher	11	5.50
5.	Winnower	34	17.00
6.	Sprinkler	0	0
7.	Sprayer	117	58.50
8.	Mould board (MB) plough	165	82.5

It is observed that 6 numbers (3%) of small farmers have tractors, 18 numbers (9%) have power tiller while as much as 117 numbers (58.5%) have sprayers and 34 numbers (17%) have winnower in the area under study. Thresher and M.B. plough found to be with 5.5% and 82.5% of the sample respectively. The inference of the Table 35 is that the costly farm machinery is hardly possessed by the sample under study for the reason of cost and skill to operate.

5.5. Communication behaviour

Mass media are the major sources of information for the farmers in village areas. All the mass media now have specific time and space for farm information. The present observation reveals that small farmers mostly use radio, TV and news paper to gather technical information about the farm practices they undertake. The relation between use of mass media and level of perception of the respondents appear in Table 36.

Analysis reveals that 92% of the sample is well exposed to different mass media like newspaper, TV and radio, against 8% who are not well exposed. Further, the respondents of medium level of perception are more exposed to mass media than those of low perception group. Chi square value being significant confirms to the findings of others that mass media regulates the formation of perception about sustainability of agriculture to a significant level.

5.6. Marketing behavior

Marketing is now an integral part in taking decisions in farming. Irrespective of level of farming market is an important input to sustain the level of productivity and income of the farmers. The present study included three important components namely, marketable produce, opinion about sale price and place of market of farm produce.

Table 36: Exposure to mass media and perception of sustainability in agriculture

(n=200)

Sl.No	Exposure to mass media	Low		Medium		High		Total		χ^2 Value
		F	%	*f*	%	*f*	%	*f*	%	
1.	Exposed	34	80.95	102	94.44	48	96.00	184	92.00	8.929*
2.	Not exposed	8	19.05	6	5.56	2	4.00	16	8.00	
3.	Total	42	100	108	100	50	100	200	100	

* Significant at 5% level of probability

i. Marketable produce

Small farmers possess small land holdings but make intensive use of it. In many cases the Cropping intensity in coastal area of Odisha is more than 300% in case of small farmers. Farmers grow more cash crops beside cereal to sell and earn living. In analyzing the opinion of the respondents about marketable farm produce at house hold level the following information were obtained.

Data reveal that about 80% of sample have little surplus over the family requirements to sell in market, where as 10.5% mentioned that their produce is just manageable. About 9.5% of them mentioned their farm produce is more than family requirement. The χ^2 -test is not significant as the sample is homogenous in land holding and production level.

ii. Opinion about sale price

Sale of farm produces on remunerative price act as parameter for assessing the sustainable production and productivity. Many of the small farm operators report that whatever produce they sell in the market do not compensate to cost of production. In many parts of the state more production has become a problem with farmers because of non-existence of regulated market. In examining the opinion of the respondents about remunerativeness of farm produce, the information received are presented in Table 38.

A majority of the sample (54.5%) expressed that farm produce is not remunerative while 25.5% mentioned as remunerative against 8.5% who are less concerned about the sale price. Farm produce as quite remunerative was mentioned by 11.5% sample. The Chi square value reveals a significant relationship between perception and opinion about sale price among the sample under study.

iii. Place of marketing

The disposal of farm produce is an important aspect in sustainability in agriculture. A number of studies have indicated adoption of technology and place of marketing of farm produce are interrelated. In examining the place of market of farm produce under the sub items of own locality, outside locality and outside district in the state, the results obtained are presented in Table 39.

Findings reveal that only 3.5% of the sample farmers of medium and high level of perception could establish market links outside the district where as 59% sell out side of own locality and 37.5% in own localities. The farm produce being small quantities, the distant market link is probably not required for which such results have been obtained. As hypothesized, place of marketing of farm produce is related to the level of perception of small farmers (χ^2 -test significant).

Table 37: Marketable farm produce and perception of sustainability in agriculture

(n=200)

Sl.No	Farm produce	Low		Medium		High		Total		χ^2 Value
		F	%	*f*	%	*f*	%	*f*	%	
1.	Just manageable	8	19.03	11	10.19	2	4.00	21	10.50	6.569N.S.
2.	Little over family maintenance		32	76.19	85	78.70	43	86.00	160	80.00
3.	Much more than family requirement		2	4.76	12	11.11	5	10.00	19	9.50
	Total	42	100	108	100	50	100	200	100	

N.S-Non Significant

Table 38: Opinion about sale price and perception of sustainability in agriculture

(n=200)

Sl.No	Opinion	Low		Medium		High		Total		χ^2 Value
		F	%	*f*	%	*f*	%	*f*	%	
1.	Remunerative at times	12	28.57	29	26.85	10	20.00	51	25.50	22.404*
2.	Quite remunerative	0	0	17	15.74	6	12.00	23	11.50	
3.	Not remunerative	21	50.00	54	50.00	34	68.00	109	54.50	
4.	Less concern	9	21.43	8	7.41	0	0	17	8.50	
5.	Total	42	100	108	100	50	100	200	100	

* Significant at 5% level of probability

Table 39: Place of marketing and perception of sustainability in agriculture (n=200)

Sl.No	Market place	Low		Medium		High		Total		χ^2 Value
		F	%	*f*	%	*f*	%	*f*	%	
1.	Own locality	18	42.86	27	25.00	30	60.00	75	37.50	25.497*
2.	Outside locality	24	57.14	78	72.22	16	32.00	118	59.00	
3.	Outside district	0	0	3	2.78	4	8.00	7	3.50	
4.	Total	42	100	108	100	50	100	200	100	

* Significant at 5% level of probability

5.7. Farming behaviour

Farming experience and responsibilities at farm level are almost same with all the respondents selected for study. They being the small farmers possessing equal size of farm growing similar type of crops and residing under similar eco-system are almost same in many aspects. Since all the farmers take farming as occupation and involve directly in farming process, the farming experience and responsibilities could not be identified as parameters to bring difference in perception level.

5.8. Training

Training is another important component of present farming system in view of advancement in science and technology. It has been a policy of Government. to provide training to farmers with evolution of new technology in the field of agriculture. The training availed by the sample under study is reflected in Table 40.

Of the sample, 80.5% have availed training against 19.5% who did not. The frequency of training is more pronounced in case of high level of perception followed by medium and low level. The Chi square value being significant (17.828), the relationship between perception and training undergone is found to be interrelated. In short the socio-economic and farm related variables are presented to have over all glance of the objective.

Table 40: Training undergone and perception of sustainability in agriculture

(n=200)

Sl.No	Training undergone	Low		Medium		High		Total		χ^2 Value
		F	%	*f*	%	*f*	%	*f*	%	
1.	Yes	25	59.52	89	82.41	47	94.00	161	80.50	17.828*
2.	No	17	40.48	19	17.59	3	6.0	39	19.50	
3.	Total	42	100	108	100	50	100	200	100	

* Significant at 5% level of probability

Table 41: Perceived attributes of sustainability and socio-economic profile

Sl No	Variables	χ^2 – Value	Significant (S)/ Non- significant (NS)
1.	Age	13.05	S
2.	Educational status	43.32	NS
3.	Family composition	1.645	NS
4.	Family type	191.45	S
5.	Caste	23.98	S
6.	Membership in formal organization	24.18	S
7.	Land possession	14.3587	NS
8.	Occupation	1.787	NS
9.	Crops	3.026	NS
10.	Animal wealth	0.396	NS
11.	Additional source of income	3.703	NS
12.	Use of energy	11.398	S
13.	Farm machinery	-	-
14.	Exposure to mass media	8.929	S
15.	Marketable farm produce	6.569	NS
16.	Opinion about sale price	22.404	S
17.	Place of marketing	25.497	S
18.	Farming behaviour	-	-
19.	Training	17.828	S

5.9. Variation at district level

The sample was drawn from coastal districts of Puri (D1), Khurda (D2), Ganjam (D3) and Balasore (D4). The variation among the sample of different districts have been calculated and presented below.

Table 42: Variation in the profile of sample of districts under study

Sl.No	Profile of sample	Districts				
		D1	D2	D3	D4	Average
1. Personal						
i	Age	48	49.5	49.5	50	49.2
ii	Education	2.50[a]	2.06[b]	1.96[b]	1.82[b]	2.08
2. Social						
i	Family composition	5.54[a]	6.14	6.30[b]	6.70[b]	6.17
ii	Family type	1.48	1.44[a]	1.50[b]	1.46	1.47
iii	Caste	3.30[a]	3.42[b]	3.34[a]	3.34[a]	3.35
iv	Membership in formal organization	0.94[a]	1.04[a]	1.92[b]	1.64	1.39

(Contd.)

Sl.No	Profile of sample	Districts				
		D1	D2	D3	D4	Average
3. Economic						
i	Land holding	3.18^{a}	3.58^{b}	3.30^{a}	3.28^{a}	3.34
ii	Occupation	2.20^{a}	2.42^{b}	2.28	2.22^{a}	2.28
iii	Irrigation	2.58^{a}	2.32	2.52^{a}	1.89^{b}	2.33
iv	Crops grown	1.84^{a}	2.30^{b}	2.00	2.22^{b}	2.09
v	Animal wealth	1.22^{a}	1.28^{x}	1.30^{b}	1.22^{ay}	1.26
vi	Additional source of income	2.36^{a}	2.24^{b}	2.38^{a}	2.52	2.38
vii	Use of energy	1.26	1.24^{a}	1.34^{a}	1.14^{b}	1.25
viii	Farm implements	1.67^{a}	2.80^{b}	2.16	1.67^{a}	2.08
4. Communication Behaviour						
i	Exposure to mass media	9.27^{a}	7.73^{bx}	8.93^{a}	9.06^{y}	8.75
5. Marketing Behaviour						
i	Place of marketing	1.25^{ax}	0.88^{b}	0.78^{b}	0.89^{y}	0.95
ii	Opinion about sale price	2.30^{a}	2.52^{b}	2.42	2.30^{a}	2.34
iii	Marketable produce	1.92^{a}	2.08^{b}	2.04	1.92^{a}	1.99
6. Farming Behaviour						
i	Experience in farming	1.92^{a}	1.94	1.96^{b}	1.96^{b}	1.95
ii	Farming responsibilities	3.10^{a}	3.20^{a}	2.98^{x}	3.68^{by}	3.24
7. Training						
i	Training undergone	1.82	1.76^{a}	1.94^{b}	1.69^{a}	1.80
ii	Use of training knowledge	1.00	1.00	1.00	1.00	1.00

a,b and x,y : different parenthesis indicate significant difference ($p<0.05$)

The findings of study revealed that:

1. Selection of sample at the district level with regard to age does not exhibit difference and hence selection of sample is appropriate.
2. With regard to education wise difference was observed among the districts of Puri, Khurda, Ganjam and Balasore.
3. In case of family composition, the district Puri exhibited a difference with Ganjam and Balasore whereas in case of family type only difference was observed within Khurda and Ganjam.
4. In case of caste, the composition of Khurda was different with that of Puri, Ganjam and Balasore.
5. Membership in organization was found to be different between Puri and Ganjam and Khurda and Ganjam.

6. In case of land holding, Puri and Khurda, Khurda and Ganjam and Khurda and Balasore have significant difference.
7. Invariably Khurda observed to be different than Puri and Balasore in case of occupation.
8. With regards to number of crops grown, Puri has significant difference with Khurda and Balasore.
9. Possession of animal wealth is found to be different among three districts of Puri and Khurda, Puri and Ganjam, Khurda and Balasore and Ganjam and Balasore.
10. Additional sources of income equally found to be different among the small farmers of Puri and Khurda, Ganjam and Balasore.
11. District level variation was observed in case of Khurda and Balasore as well as Ganjam and Balasore in case of energy use.
12. Possession of farm machinery was different with sample of Puri and Khurda as well as Khurda and Balasore.

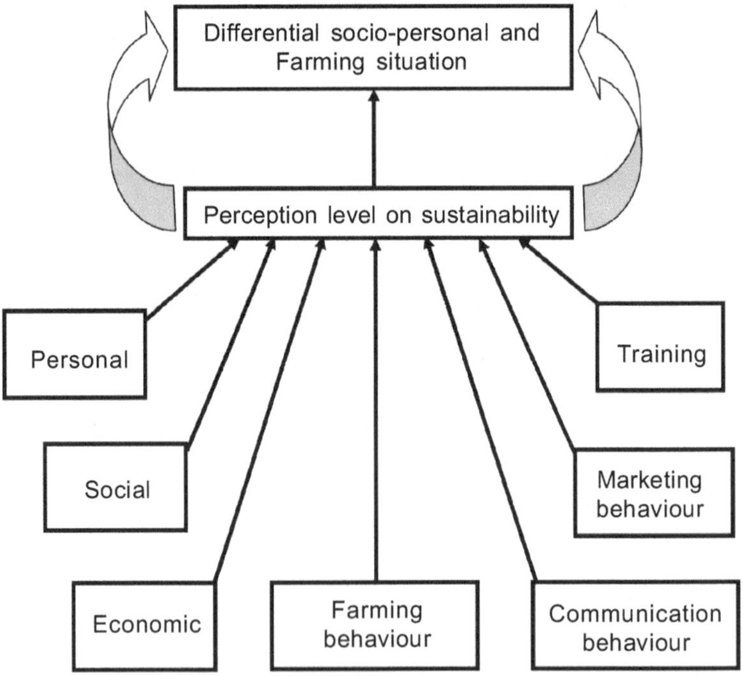

Fig. 3: Relationship between profile of sample and level of perception

13. Absorption of information from mass media was found to be different among the sample of Puri and Khurda, Khurda and Ganjam as well as Khurda and Balasore.
14. Similarly place of marketing was significantly different in case of Puri, Khurda, Ganjam and Balasore.
15. The experience of farmers about sale price reveals difference among the farmers of Khurda and between Puri and Balasore.
16. Marketable farm produce was differently viewed by the farmers of Puri, Khurda and Balasore districts.
17. Difference in training undergone is found in case of Ganjam, Khurda and Balasore.

6

Organic Farming and Small Farm Operating System

Organic farm produce is now gaining momentum in the world. People of high status and concerned with health hazards are interested in organic product. Since 1990, the market of organic product has grown reaching a value of Rs 55 billion in 2009 as per the Organic Monitor in the world. The demand has been reflected in increase of land under organic farming which has grown correspondingly at a rate of 8.9% per annum. At present there is approximately 37,000,000 hectare of land under organic farming worldwide. The idea has been percolated to rural areas and the farmers of all kinds are aware of organic farming. In terms of organic land, Australia leads the pack with 11.3 million hectares followed by Argentina (2.8 million hectares) and Italy with more than one million hectares. In Asia, China, India and Japan are the largest organic producers. Japan is the region's largest market and buys majority of Asia's production.

6.1. Organic farming

Organic farming is a form of agriculture that relies on techniques such as crop rotation, green manure, and biological pest control. It uses fertilizers and pesticides other than synthetic chemicals. Organic farming makes no use of plant growth regulators such as hormones, livestock antibiotics, food additives, genetically modified organisms, human sewage, sludge and none materials.

Organic farming is managing the agro-ecosystem as an autonomous system based on primary production capacity of soil under the local condition. Agro-ecosystem management implies treating the system, on any scale as a living organism supporting its own potentials for biomass and animal production along with biological mechanism for mineral balancing, soil improvement and pest control. It is a system of farm design and management to create an ecosystem which can achieve sustainable productivity without use of artificial external inputs such as chemical fertilizer and pesticides. Lampkin (1990) defines organic farming as a production system which avoids or largely excludes the use of synthetic compound fertilizers, pesticides, growth regulators and livestock feed additives. To the maximum extent possible, organic farming concentrates in crop rotation, residues, manures, legumes, green leaves manures, off farm organic wastes and biological methods of pest control.

6.2. Brief history of organic farming

The original agriculture was organic farming and has been practiced for thousands of years. Forest gardening is taken as fully organic production system and is thought to be world's oldest agro ecosystem. Organic movement began in mid-1920 in central Europe by Rudolf Steinger. Organic farming was independently developed in 1940s in England through the work of Albert Howard. Artificial fertilizer was developed in 18th century initially with super phosphate followed by ammonia based fertilizers. Albert Howard is considered as father of organic farming as he was the key founder during post-industrial-revolution organic movement. Further work was done in USA and other countries in the world.

6.3. Objectives of organic farming

The concept of organic farming was developed with following objectives:

1. Farming on local resources
2. Maintaining soil fertility
3. Avoiding pollution caused by agriculture
4. Obtaining food stuffs of high quality and sufficient quantities
5. Reducing use of fossil energy
6. Meeting physiological needs of livestock
7. Development of potentiality of farmers through work
8. Maintaining rural environment
9. Preserving non-agricultural ecological hub.

6.4. Components of organic farming system

i. Organic manure

Farm Yard Manure (FYM), compost, biofertilisers, green manures, biogas slurry, cover crops and green manuring crops, straw and other crop residues, other agricultural and industrial wastes etc. suitable for composting can substitute for inorganic fertilizers to maintain soil health and environmental quality while producing optimum crop yield. Sea weeds, fish manures, blood and bone meals and some permitted fertilizers like basic slag and rock phosphates can be used in organic farming.

Organic matter contents of the soil will be increased by use of organic manures. This in turn increases the water holding capacity, physical, chemical and microbial properties of the soils and ultimately the crop yields. Erosion is reduced by use of organic manures. Soil fertility increases by crop rotation with legumes. Green manuring provides nutrients and brings improvement in soil.

ii. Waste recycling

By way of composting organic waste are converted into organic fertilizers by means of biological activities under controlled condition. The recycling of organic waste materials as manures for maintaining and sustaining soil fertility is very important for organic farming. The farm house hold and cattle shed wastes, vegetable and animal refuses (rural and urban), cereal straws, stubbles, cotton stalks, ground nut husks, farm weed and grasses, leaves, leaf moulds, wood ash, litter, urine soaked earth from cattle shed and such other substances are composted by various methods like NADEP, Coimbatore method, Bangalore method and Indore method etc.

Vermicomposting is one of the most natural, ancient and supposed to be best among all sustainable practices. Vermicompost is the organic manure produced by earth worms which is a mixture of worms and casting, organic materials including humus, live earth worms, their cocoons and other micro organisms. Kitchen wastes, city wastes, sewage wastes etc. can be utilized for making vermin compost.

iii. Non chemical weed control

In case of organic farming, crop cultivation is done without use of chemicals for weed control. Farmers apply only naturals and no chemical herbicides are used. This prevent environmental pollution.

iv. Biological pest management

Control of insect pest and diseases is the major challenges of tropical and subtropical farming.

In organic farming the non chemical and biological pest control and management is followed. The natural enemies of pests are conserved. This is important for minimizing the use of chemical pesticides and avoiding multiplication of insect and pest resistant to pesticides. Botanical pesticides and selected microbial pesticides like *Bacillus thuringiensis* are promising pesticides for organic farming.

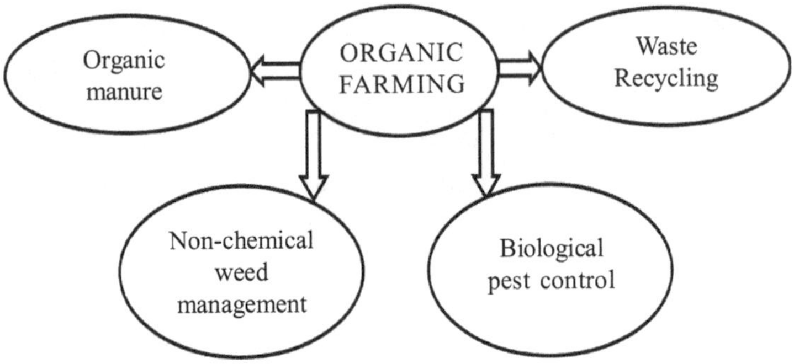

6.5. Concept of organic farming

Concept of organic farming is as old as agriculture on earth. Many scientists have defined the concepts taking local environment into account. Some of such views are cited here to have better understanding of the concept.

- Organic farming produces food and fiber without use of toxic chemicals and depends on biological diversity.
- Organic farming substitutes inorganic fertilizers with manure and other organic matters.
- It makes use of biological pest control methods.
- It is a conventional method of agriculture.
- It has more relevance to backyard gardens.

Organic agriculture is an ecological production management system that promotes and enhances biodiversity, biological cycles and soil biological activity. It is based on minimal use of off-farm inputs and management practices that restore, maintain and enhance ecological harmony.

6.6. Need of organic farming

People are becoming aware of organic farming throughout the world. The need of organic farming at present is based on the following considerations.

a) Organic farming helps in finding a path to establish harmony with nature.

b) For sustainable environment, organic farming is a must.

c) Organic farming minimizes environmental pollution.

d) It checks soil erosion and conserves soil fertility.

e) It is best suited for limited land holdings and pro-poor economic condition.

f) Food grains produced through organic farming does not contain any toxic materials.

6.7. Present status of organic farming

Organic farming is now practiced in almost all the countries of the world. According to SOEL-FIBL Survey (2007) the total geographically managed area is more than 30 million hectares in the world.

Table 43: Organic land and farms by continents

Sl. No.	Continents	Organic land (ha.)	Organic farms
1.	Asia	2,893.372	120,805
2.	Africa	809.504	124,805
3.	Europe	6,920.462	187,697
4	Latin America	5,809.32	176,710
5.	North America	2,199.225	12,063
6.	Oceanica	11,845.100	2,689
	Total	30,558.183	633,891

Source: SOEL-FIBL Survey 2007.

Area under organic certification process which observed a rapid growth till the year 2008-09 had shown a declining trend till 2010-11.

Table 44: The growth of area under organic management

S. N.	Year	Area in ha under organic certification	
		Cultivated (organic + in-conversion)	Wild harvest
1.	2003-04	42000	NA
2.	2004-05	76000	NA
3.	2005-06	173000	NA
4.	2006-07	538000	2432500
5.	2007-08	865000	2432500
6.	2008-09	1207000	3055000
7.	2009-10	1085648	336000
8.	2010-11	777517	3650000

Both certified and non-certified streams of organic farming are growing. Continuous organic projects are also opting in or out third party certification system depending upon the need of certification in direct marketing system.

Table 45: Total area under organic certification process and total production of raw organic products (cultivated + wild harvest) in India during 2010-11

Sl.No.	State	Area (M ha)	Production (MT)
1.	Andhra Pradesh	14350.62	59470.76
2.	Arunachal Pradesh	243.1	2127.29
3.	Assam	2047.09	14716.95
4.	Andaman	334.68	4189.10
5.	Bihar	1303.62	15153.35
6.	Chhattisgarh	8448.93	1695.82
7.	Daman Diu		
8.	Delhi	266.32	2172.26
9.	Goa	13303.32	28262.5
10.	Gujarat	48518.1	191667.84
11.	Haryana	14763.61	119789.39
12.	Himachal Pradesh	631901.99	74973.30
13.	J&K	776.48	10382.95
14.	Jharkhand	24300	0
15.	Karnataka	88728.64	220901.31
16.	Kerala	6597.65	58177.29
17.	Lakshadweep	12.13	22.55
18.	Manipur	2792.03	19239.25
19.	Maharashtra	177345.48	694275.26
20.	Madhaya Pradesh (MP)	2866571.88	122080.58
21.	Mizoram	12544.13	177509.02
22.	Meghalaya	2419.67	15674.64
23.	Nagaland	1603.54	6627.47

(Contd.)

Sl.No.	State	Area (M ha)	Production (MT)
24.	Odisha	24417.55	166183.41
25.	Punjab	6025.78	68177.83
26.	Rajasthan	217712.19	265341.01
27.	Sikkim	1726.34	5174.44
28.	Tamilnadu	34878.09	41640.73
29.	Tripura	348.39	527.25
30.	Uttar Pradesh	111644.83	294156.10
31.	Uttarakhand	105465.98	79765.04
32.	West Bengal	6125.72	28393.48
	Total	4427519.06	3887197.19

6.8. Characteristics of organic farming

While organic adoption may require more than refraining from the use of synthetic agrochemicals, it also offers many other benefits beyond the simplistic chemical-free caricature. The more successful farmers appear to convert for several reasons that go beyond earning a higher price for their crops. While income considerations are predominant, there are at least five major reasons why farmers choose to adopt organic practices:

a) To earn more for their production

b) To reduce or eliminate the need for purchased inputs

c) To avoid potentially harmful agrochemicals

d) To reduce their risks through crop diversification and improved soil quality/ stability

e) To maintain or improve valuable local natural resources and biodiversity.

Further, the other characteristics of organic farming include:

1. Maintains fertility of soil
2. Maintains organic content of soil
3. Provides crop nutrients for action of soil microorganism
4. Nitrogen sufficiency through legumes
5. Crop rotation prevents weeds, pests and diseases
6. Livestock management becomes cheaper
7. Farmers escape many of disasters.

6.9. International Federation of Organic Agriculture Movement (IFOAM) (1998)

We can assume that a future is both ecologically sound and economically sustainable through organic agriculture alone. In this direction the basic standards of Organic Farming and Food Processing were being set by International Federation of Organic Agriculture Movement (IFOAM) during the year, 1998.

IFOAM has stressed upon the following principles for organic agriculture:

a) To produce food of high nutritional quality and in sufficient quantity.

b) To work with natural system rather than seeking to dominate them.

c) To encourage and enhance the biological cycles within farming system involving microorganisms, soil flora and fauna, plants and animals.

d) To maintain and increase long term fertility of soil.

e) To use so far as possible renewable resources in locality through organic agricultural system.

f) To work as much as possible within a closed system with regard to organic matter and nutrient elements.

g) To give all livestock, condition of live that allow them to perform all aspects of their innate behavior.

h) To avoid all forms of pollution that may result from agricultural techniques.

i) To maintain the genetic diversity of the agricultural system and its surroundings including the protection of plants and wild life habitats.

j) To allow agricultural producers an adequate return and satisfaction from their work including a safe working environment.

k) To consider the wider social and ecological impact of the farming system.

6.10. Advantages of organic farming

The various benefits of organic farming for small farmers all over the world include a high premium, low capital investment, the ability to achieve higher premiums in the market, and the ability to use traditional knowledge. According to research conducted by the Office of Evaluation and Studies (OE), at the International Fund for Agriculture Development (IFAD), small farmers in Latin America, China, and India can benefit dramatically from organic farming and this will help in alleviating poverty in these countries.

Organic farming refers to means of farming that does not involve the use of chemicals such as chemical fertilizers and chemical pesticides. Numerous small farmers practice organic farming; however, since they are unaware of the market opportunities, they are not able to reap the benefits of organic farming. Some of the advantages of organic farming for small farmers are discussed:

a) **High premium:** Organic food is normally priced 20-30% higher than conventional food. This premium is very important for a small farmer whose income might only be sufficient to feed his/her family with one meal.

b) **Low investment**: Organic farming normally does not involve capital investment as high as that required in chemical farming. Furthermore, since organic fertilizers and pesticides can be produced locally, the yearly costs incurred by the farmer are also low. Agriculture greatly depends on external factors such as climate, pests, and disease. Furthermore, most of the small farmers are dependent on natural rain for water. Therefore, in cases of natural calamity, pest or disease attack, or irregular rainfall, when there is a crop failure, small farmers practicing organic farming have to suffer less as their investments are low. It should be noted that while shifting from chemical farming to organic farming, the transition might be costly.

c) **Less dependence on money lenders:** Many small farmers worldwide commit suicide or suffer from intense economic strife due to increasing debt. Since chemical inputs, which are very costly, are not required in organic farming, small farmers are not as dependent on money lenders. Crop failure, therefore, does not lead an organic farmer into enormous debt, and does not force him to take any extreme steps.

d) **Synergy with life forms:** Organic farming involves synergy with various plant and animal life forms. Small farmers are able to understand this synergy easily and therefore find it easy to implement the organic farming techniques.

e) **Traditional knowledge:** Small farmers have an abundance of traditional knowledge with them and within their community. Most of this traditional knowledge cannot be used for chemical farming. However, when it comes to organic farming, farmers can make use of this traditional knowledge. Furthermore, in cases of organic farming, small farmers are not dependent on those who provide chemical know-how.

In general, the organic farming has the following advantages.

I. Organic manures produce optimal conditions for high yield and good quality crop.

II. Organic manures supply all required nutrients of crop.

III. Improves plant growth and physiological activities of plant.

IV. Improves soil physical properties such as good tilth, good aeration, easy root penetration and thus improving water holding capacity.

V. Improves soil chemical properties such as retention of soil nutrients.

VI. Helps to prevent environmental degradation and can be used to regenerate degraded areas.

VII. Organically grown crops are more resistant to disease and pest attack.

VIII. It provides superior food for human beings.

IX. Consumer demand for good food is taken care of by organic farming.

6.11. Climate change and agriculture

Climate change is defined as change in climate over time, whether due to natural variability or as a result of human activity. Adaptive capacity is the ability of a system to adjust to climate change (including climate variability and extremes) to moderate potential damages, to take advantage of opportunities and to cope with the consequences.

Vulnerability is the degree to which a system is susceptible to and unable to cope with adverse effects of climate change including climate variability and extremes. New option for carbon sequestration in agriculture, forestry and land-use must change as deforestation contributes to respectively, 13 and 17 percent of total anthropogenic greenhouse gas emissions while carbon dioxide emission from agriculture is small. The sector account for about 60 percent of all nitrous oxide (N_2O, mainly from fertilizer use) and about 50 percent of methane (CH_4), emitted mainly from natural and cultivated wetlands and enteric fermentation. The Inter Governmental Panel of Climate change (IPCC) estimates that the global technical mitigation potential for agriculture (excluding forestry) will be between 5500 and 6600 mt CO_2- equivalent per year by 2030, 89 percent of which are assumed to be from carbon sequestration in soils.

Climate change has significant impact on agriculture and specifically on crop growth. Impact on crop growth affects farmers and their farm enterprise. Subsequently the climate brings a remarkable change in life style of the farmers.

The crop growth in respect of physiology, phenology and morphology are directly affected by climate change.

Agriculture sector contributes 14% of global greenhouse gasses of which 74% comes from developing countries. There is around 70% potential in developing countries to mitigate this effect through effective farming practices which capture carbon and store it in soils that is deemed to be cost effective as well as preventive measures that can be undertaken in developing countries. The contribution of Indian agriculture to climate change is well established. Agricultural practices like, flooding of paddy field, addition of nitrogen oxide through fertilizers, combustion of fossil fuels etc. add to the emission of green house gases. Nitrous oxide is more dangerous than carbon dioxide with global warming potential 296 times greater than carbon dioxide. It is estimated that 28% of the GHG emissions and about 78% of methane and nitrous oxide emissions are from agriculture. The Inter Governmental Panel on Climate Change (IPCC) has estimated that every quintal of nitrogen applied in farming emits 1.25kg of nitrous oxide and globally half of the nitrogen applied to crop is lost due to the environment. Burning of crop contributes to GHG emission. The crop residue from every 4 tons of rice or wheat grain is about 6 tons of straw. Burning of crop residue also adversely affect soil fertility.

Climate change can impact agricultural sustainability in three interrelated ways: first, diminishing the long-term ability of agro-ecosystems to provide food and fiber for the world's population; and second, by inducing shifts in agriculture regions that may encroach upon natural habitats at the expense of floral and faunal diversity. Global warming may encourage the expansion of agricultural activities into regions now occupied by natural ecosystem such as forests, particularly at mid-and high-latitudes. Forced encroachments of this sort may thwart the process of natural selection of climatically-adapted native crops and other species while the over-all global impact of climate change on agricultural production may be small, regional vulnerabilities to food deficits may increase due to problems of distribution and marketing food to specific regions and group of people. For subsistence farmers and more so poor people who now face a shortage of food, lower yields may result not only in measurable economic losses but also in malnutrition and even famine. On the biophysical side, due to cultivation of C_3 crops, temperature is likely to be more responsive to increasing levels of CO_2. Tropical crops are closer to their high temperature optima and experience high temperature stress despite lower projected amounts of warming. Insects and diseases, already much more prevalent in warmer and more humid regions may become even more widespread.

a) Impact of climate change

In India the direct impact of climate change would affect plant growth, development and yield due to change in rainfall and temperature. Increase in temperature would reduce crop duration, increase crop respiration rate, change the pattern of pest attack, a new equilibrium between crop and pest thus observed hasten mineralization in soil and decrease fertilizaer use efficiency. All these could considerably affect crop yield for long run. In general the simulation results indicate increasing temperature, with decreasing growth and yield of agricultural crops. Increased CO_2 levels are expected to favor growth and increase crop yields and therefore will be helpful in counteracting the adverse effects of temperature rise in future. On global level climate change effects will change the crop production areas. In middle and higher latitudes, global warming will extend the length of the potential growing season allowing earlier planting of crops in the spring, earlier maturation and harvesting and the possibility of completing two or more cropping cycles during the same season.

Crop-production areas may expand pole ward in countries although yields in higher latitudes will likely be lower due to climate change. Many crops have become adapted to the growing season, day lengths of the middle and lower latitudes and may not respond well to the much longer days of the higher summers. In warmer lower latitude regions increased temperate may accelerate the rate at which plant release CO_2- in the process of respiration resulting in hastened maturation and reduced yield.

Effects of global warming in India

Global warming results to

1. Increased temperature
2. Effect on monsoon
3. Effect on water resource
4. Effects on agriculture
5, Human health
6. Effects on forest
7. Effect on coastal land and delta.

6.12. Management of organic farming

Farmers are well acquainted with organic farming which they follow from generation to generation. With advancement of agricultural techniques and use of new varieties, they have opted to use chemicals which now have given rise to numerous problems. The management of organic farming is more important to keep productivity at desired level and at the same time the environment becomes sustainable. In other words, it is known as management of autonomous ecosystem cent per cent. Organic US rule regulating the organic product is as follows.

Organic means	=	more than 95% organic ingredients
Made with organic ingredients means	=	70-95% organic ingredients

6.13. Initiating organic farming

Organic farming is most suitable in dry land farming because dry land provides suitable ground to opt for organic farming. The advantages that the dry land offers for organic farming are:

a) It contains comparatively less residual effect of chemical fertilizers and pesticides.

b) In dryland, effect of organic manure and water holding capacity is more and quickly visible than irrigated land.

c) Broad Methods of organic farming.

Broad methods of organic farming

1. Soil Management
2. Weed Management
3. Controlling other organisms
4. Genetic modification

6.14. Methods followed in management of autonomous ecosystem

Autonomous ecosystem management is the prime requirement for sustainable ecosystem and thereby bringing sustainability of agriculture. Important methods of managing sustainable ecosystem are discussed.

1. Mixed farming
2. Plants
3. Animals
4. Soils

5. Biosphere
6. Crop rotation
7. Crop diversification
8. Organic cycle optimization
9. Partnership with nature

1. Mixed farming

Mixed farming is very common among Indian farmers. Crop and livestock in a farm promote agro-ecosystem because of cyclic process and integration resulting crop residue and animal based manure. This is an excellent opportunity to opt for organic farming. Diversification of species, biotypes and land use are means to optimize agro-ecosystem. The concept is quite old and was popular in ancient India.

2. Plant

The soil, animal and plants have multifunctional activities in farm. They produce human foods herbs and medicines, non food product like fuel, flower, fiber etc. and animal feeds.

3. Animals

Domestic animals within an eco-system produce multiple benefits. They produce protein, feed and manure to provide nutrients to crops. The ruminants are essential for organic farming. Domestic animals are linked with the soil resulting in feed and manure production, livestock health for good quality meats, appropriate breeds, proper housing, feed and hygiene. In short animals constitute major elements of an organic farming.

4. Soil

Living soils are necessary for organic farming. The deep rooting plants have the capacity to bring mineral deposits from deeper soil layers to nutrient cycle. The activities like nodulation, N- fixation by legumes, N-balance contribution, denitrifying bacteria, P-solubilising bacteria and root symbiosis with mycorrhiza are all highly essential for crop growth. The problems like soil degradation and soil erosion are constant problems in farming. Organic farming looks after efficiency in use of resource and energy and quality of food. The basis of organic farming is maximization of biological effects and minimization of external inputs. The concept as such has been accepted throughout the globe and industrial product at present is being directed towards organic content.

5. Biosphere

The local biosphere with minimal external inputs and maximal use of ecological potential provides a strong base for organic farming. Biosphere provides interactive synergy of crop and animal, rotation and manure to comply with local and seasonal conditions. Microclimate is affected by trees and shrubs, irrigation, adaptation of varieties and many other factors. The logic behind is direct radiation from sun on soil surface is prevented by using soil covers and mulches. Thus keeping biosphere in harmony with our crop system can lead to organic farming.

6. Crop rotation

Crop rotation is a must in organic farming. Cultivating a wide variety of different crops in a specific sequence is called crop rotation. The benefits of inter-cropping, mixed cropping and sowing in different conditions provide good ground for organic farming.

7. Crop diversification

Crop diversification is another indicator of organic farming. The benefits of crop diversification are yield stability, reduction of disease and pest, weed control, control of soil erosion, recycling of nutrients and transfer of N from N-fixing species. Crop rotation and multiple cropping systems have structural advantages like sowing at staggered intervals, crop maintenance and reduced labor peaks and harvest.

8. Organic cycle optimization

Each plot and each farm maintains nutrients. Nutrients are to be used in balance. Optimal use means nutrients should be recycled and used in various times. The principle says that the minimum amount of nutrient is to be maintained in soil. The quantity of nutrients available to plants and animals can be increased by activating the edaphon resulting in increased weathering of mother materials. Quality of biomass produced and incorporated in recycling process adds to organic matter. Recycling heals up the gaps through physical and biological means making system fit for organic farming.

9. Partnership with nature

Some authors are in view on organic system that a holistic farming system can be developed with autonomous agro-ecosystem partnership with nature. The system being static, changing condition by use of practical experience and knowledge is necessary. But it requires attributes which emphasizes human in ecosystem capacities.

6.15. From conventional to organic farming

In local situation organic farming system cannot be introduced in a short period as farmers are practicing inorganic farming for a long period. The conversion period denotes change from inorganic to organic system. It is a matter of coordination to examine the of traditional farming, area covered under inorganic farming and constituent of soil best managed to fulfill the requirement of organic farming. The composition of farm widely varies according to geographical conditions, ownership structure, time span and related factors. From conventional to organic farming is a time taking process.

In addition to above practices, Desai and Poojari (2007) classify approach to sustainable agriculture under LEISA (Low External Input Sustainable Agriculture) as biological husbandry, biodynamic farming, organic farming, perma-culture, vedic agriculture and natural farming.

The practices recommended

I. Biological practices:

a) Green manuring
b) Cover cropping
c) Composting
d) Companion planting
e) Integration of crop and livestock
f) Tillage and cultivation

II. Biodynamic practices:

a) Special compost preparation
b) Special foliar spray
c) Planting by calendar
d) Peppering for pest control
e) Homoeopathy
f) Radionics

6.16. Farmers' awareness about organic farming

Small farm operators view organic farming from the point of their situations, knowledge and local feeling. Unless these farmers are well aware of organic farming and its usefulness, it becomes much more difficult to go with it. The sample farmers were exposed to organic farming in terms of its essential components, benefits, knowledge of practice, problems and suggestions for spread of organic farming.

1. Awareness about the essential components of organic farming

Farmers in general are practicing farming using chemical fertilizers and pesticides for which all out efforts were made. Now to shift it to organic farming, by changing mindset of the farmers is difficult. Before making such attempt, it is necessary to know their level of awareness about the components of organic farming. Examining the issue with sample respondents who are small farm operators, the results obtained were presented in Table 46. Their response was measured in three point scale in terms of very much, much and not aware with assigned score of 3, 2 and 1 respectively.

Table 46: Awareness about components of organic farming (N=200)

Statements	Average Mean Score	Rank
1. Non –use of chemical fertilizers and pesticides	2.10	I
2. Use of local resources	1.60	IX
3. Preservation of soil fertility	1.90	III
4. Pollution free agriculture	1.26	X
5. Production of qualitative food grains	1.70	V
6. Adoption of minimum tillage	1.62	VIII
7. Bringing stability in yield of crops	1.65	VI
8. Cost effective/less cost involved	1.63	VII
9. Consumer satisfaction	1.74	IV
10. Sound ecological environment	1.98	II

Our sample is consisting of small and marginal farmers drawn from four districts of different agro-climatic situations. The score analysis reveals that farmers are aware of non-use of chemical fertilizers in organic farming. The other components recognized by them are sound ecological environment, preservation of soil fertility, consumer satisfaction and production of qualitative food grains. They are also aware of stability of yield factor, less cost of production, and adoption of minimum tillage. In spite of awareness, the adoption of organic farming is far from expectation.

2. Knowledge about benefits of organic farming

The normal interest of the farmers is to know what the benefits are when they opt for adoption. Although organic farming is not new to the farmers, the idea is how to get more return. There are multiple benefits of organic farming. The reaction of the sample about such benefits is presented in Table 47.

Table 47: Knowledge about benefits of organic farming

Benefits	Mentions	Percentage
1.Improves physical properties of soil	105	52.90
2.More use of insoluble inputs	35	17.50
3.N-sufficiency in farm	40	20.00
4.Cotrol of weeds	58	29.00
5.Control of disease and pest	60	30.00
6.Environment improvement	90	45.00
7.Crops of good growth	98	49.00
8.Better physiological activities of crop	41	20.50
9.Better retention of soil moisture	70	35.00
10.Reduction in cost of inputs	120	60.00
11.Better use of farm wastes	140	70.00
12.Consumer preference for organic food	58	49.00
13. Better preservation quality of produce	85	42.50
14.Preventation of environment degradation	130	75.00
15.Checking of soil erosion	80	40.00
16. Better scope for animal rearing	60	30.00

The study shows that farmers have good knowledge about the benefits of organic farming. Majority of them know the benefits in terms of better use of farm wastes, prevention of environmental degradation, reduction in cost of farming, improve in soil condition, good growth of crops, and improvement in environment. However, they have less knowledge about use of insoluble inputs, and nitrogen sufficiency in field. Although farmers of small and marginal groups have good amount of knowledge in organic farming, still all of them are in inorganic farming.

6.17. Organic farming practices

Organic farming practices are known to farmers. Some of them use one or other practices that fall in the domain of organic farming. The knowledge level of sample farmers about different practices of organic farming as has been recorded is given in Table 48.

Table 48: Knowledge about organic farming practices

Practices	Know		Don't know	
	Frequency	%	Frequency	%
1. Mixed farming	40	20.00	160	80.00
2. Growing of herbs and shrubs	50	25.00	150	75.00
3. Keeping of farm animals	130	75.00	70	25.00
4. Crop rotation	90	45.00	110	55.00
5. Use of local genes	110	55.00	90	45.00
6. None use of chemicals for soil, pest and diseases	80	40.00	120	60.00
7. Vermicompost	70	35.00	130	65.00
8. Compost making	125	62.50	75	37.50
9. Green manure	150	75.00	50	25.00
10. Growing of leguminous crops	160	80.00	40	20.00
11. Use of azolla	40	20.00	160	80.00

With advance in science and technology, there has been significant change in practices of organic farming. It is not new to our farmers. Due to over emphasis on inorganic farming and provision of incentives, the farmers of all land sizes are not willing to opt for organic farming. A good percentage of farmers have knowledge about organic farming practices like, growing of leguminous crops, green manure, keeping of farm animals and use of local varieties of crops. Very small percentage of farmers know about mixed farming and growing of herbs and shrubs to supplement organic contents of the soil. The need for popularization of organic farming practice is very much realized to boost the program.

6.18. Adoption of organic farming and associated problems

Having knowledge and awareness, the farmers are not yet adopting organic farming and all of them use inorganic fertilizers in farming. The problems are many as expressed by the sample farmers and some of these problems are given in Table 49.

Table 49: Problems in adopting organic farming

Problems	Average	Rank
1. Insufficient local resources	2.40	VII
2. Not sure of good yield	2.60	IV
3. Labor intensive	2.45	VI
4. Lack of market demand	2.35	VIII
5. Lack of technical support	2.65	III
6. Lack of organized efforts for organic farming	2.75	I
7. Absence of Govt. policy	2.72	II
8. Difficult to indentify organic areas for farming	2.57	V
9. Apprehension to lose incentives given for fertilizers etc.	2.45	VI
10. Lack of emphasis and movement for organic farming	2.75	II

Problems in adoption of organic farming are many as expressed by the farmers in rural areas. These are lack of emphasis, absence of definite policy, lack of technical support, not sure of good yield, difficulty in identifying the organic area, labor intensive and lack of market demand. Many of these problems are not beyond the capacity to solve provided a sincere attempt is made.

6.19. Suggestions for adoption of organic farming

Farmers know the problem and also solutions. The important suggestions offered by the farmers under study are presented in the Table50.

Table 50: Suggestions of farmers for adoption of organic farming

Suggestion	Mentions	Percentage
1. Incentives for organic farming	140	70.00
2. Creation of market for organic produce	180	90.00
3. Training and extension support	68	34.00
4. Formation of organic farmer groups	60	30.00
5. Involvement of rural youths	64	32.00
6. Special program by govt.	88	44.00
7. Popularization of organic farming by extension agencies	140	70.00

The suggestions to increase adoption of organic farming are, creation of market for organic produce which will be definitely costly, incentives for organic farming by the Government, popularization of programme, special effort by Government, training, technical assistance, involvement of youths and formation of organic farmer group.

7

Perception of Small Farm Operators About Sustainability of Agriculture and Gaps in Comprehending Sustainability

7.1. Sustainability in agriculture

Sustainability essentially means meeting the needs of present generation without compromising the ability of future generation to meet their own needs. The concept used by FAO to convey the meaning of sustainability necessarily rely on management and conservation of resource base in relation to orientation of technological and institutional changes to ensure satisfaction in meeting human needs. Sustainability normally covers the meaning of conservation of natural resources. The natural resources are land, water, forest and livestock. The maintenance of these resources suffers due to low or no maintenance because of absence of infrastructure and inadequate finance. Apart from these constraints, poor management ability is also responsible for degradation of natural resources. To equip the people, in general, poor managerial ability has to be upgraded through multidisciplinary program, appropriate technology and establishment of peoples' organizations. To work on sustainability as defined for maintenance and conservation, the people in general ought to develop a positive perception to work before the program is initiated. The perception directs action of individual

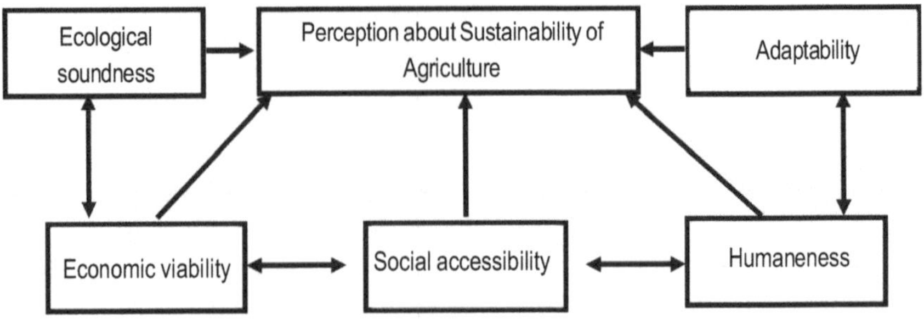

Fig. 4: Relationship among factors of sustainability in agriculture with perception

in a positive manner. Therefore, it is necessary to determine the perception of people about the concept of sustainability, particularly in the field of agriculture.

The present study after thorough review of literature and consultation with experts in line, five important factors were identified to examine the perception level of the sample (Sabarathnam, 2002). Illustration given below reveals the inter dependency of the factors at micro level to form the perception of the people about the sustainability in agriculture.

7.2. Factors of sustainability in agriculture

The factors namely ecological soundness, economic viability, social accessibility, humaneness and adaptability of technology were selected for the study to find out the relative position of these factors in the perception score about sustainability in agriculture. The ranking of factors in accordance with their relative influence on sustainability in agriculture, as rated by the respondents are presented below.

Table 51: Ranking of factors of sustainability in agriculture

Sl. No	Factors of sustainability	Low	Medium	High	Average	Rank
1.	Ecological soundness	3.36	3.50	3.70	3.52	II
2.	Economic viability	3.88	3.83	3.82	3.84	I
3.	Social accessibility	2.69	2.66	2.76	2.70	III
4.	Humaneness	2.09	2.24	2.04	2.12	V
5.	Adaptability	2.50	2.35	2.28	2.38	IV

The respondents under low perception rated economic viability followed by ecological soundness, social accessibility, adaptability and humaneness as the important factors of sustainability in order. The sample under medium level of perception viewed similarly all the factors under study. The difference was also not marked in case of high perception groups, on economic viability as followed by ecological soundness, social accessibility, adaptability and humaneness. Taking all the responses into account, the average scores reveal that economic viability tops the list followed by ecological soundness, social accessibility, adaptability and humaneness. However, in all the cases, humaneness as a factor of sustainability is ranked last indicating that in the case of development the position of humaneness is not counted much. It is further inferred that in the face of development and survival of the present, the sample ignore all other factors except economic viability to meet the requirements of the present generation. The sample under study considers economic viability as the single most important factor that decides what kind of sustainability is required to sustain the present farming system.

Next to economic viability, ecological soundness is ranked second as it covers natural resources like soil, water, forest and livestock which are life supporting systems of rural people. The social accessibility in the scale of rich and poor does not affect much as all the respondents belong to small and marginal farm families having similar accessibility to natural resources irrespective of district and state. However, the adaptability of recent recommendations to restore sustainability occupies fourth position indicating its position among the five important factors. This factor is related to economic viability as farmers weigh the profit and loss before taking up measures to sustain natural resources which form the core aspect of the study.

However, the ranking of factors of sustainability in agriculture under rank-order correlation did not reveal any significant relationship among levels of perception of low, medium and high groups. In other words, the ranking of factors of sustainability are viewed equally by the sample under study irrespective of their perception level.

i. Ecological soundness

The most important aspect of sustainability is conservation and maintenance of ecosystem. The realization at different levels about the degradation of natural resources directly or indirectly points out to the ecological component of the place. The study attempted to find out reaction of the sample regarding ecological factors that concern more to small farmers. The statements selected for recording response are given in Table 52.

Table 52: Ecological soundness and reaction of the sample

Sl.No.	Statements	Average	Rank
1.	The soil condition of an area with regard to its structure, fertility and level of organic matter reveal the soundness of the ecology	3.51	III
2.	Availability of ground water and its nearness to root zone of crops indicate good ecological status.	2.70	VI
3.	The maintenance of nutrients of soil in form of NPK and other minor elements helps to have sound ecology of an area.	2.92	V
4.	Types of farming including crop rotation, land use pattern etc are the main indicators of ecological soundness leading to sustainability.	3.31	IV
5.	Use of organic fertilizers such as FYM, compost and Green manure leads to good ecology and sustainability in agriculture	3.95	I
6.	Use of bio-pesticides and fungicides add to maintenance of ecological soundness	3.54	II
	Average	3.32	

The average score of all the statements qualifying attributes of ecological soundness is found to be 3.62. On the basis of the score, the ranking of the statements revealed that people are more concerned about "Use of organic fertilizers such as FYM, compost and Green manure leading to good ecology and sustainability in agriculture". It is so because the farmers of Odisha are in use of organic manures since from many generations. The next rank was assigned to "Use of bio-pesticides and fungicides add to maintenance of ecological soundness". The use of pesticides has invariably created a negative attitude among the farmers about its effect which have far spreading consequences. The third position was assigned to "The soil condition of an area with regards to its structure, fertility and level of organic matter content reveal the soundness of the ecology". The soil status of coastal belt has been affected due to continuous application of chemical fertilizers for which the sample perceived its condition directly related to soundness of ecology. The next position was assigned to types of farming in terms of "Types of farming including crop rotation, land use pattern etc as the main indicators of ecological soundness leading to sustainability". The farmers believe that age old practices of crop rotation, land use pattern, cultural practices not only keep the soil condition at desirable level but also add to soundness of ecology.

The availability of ground water and its nearness to root zone of crops was found to be sixth indicator of identifying different aspects of ecology. The farmers under study realize that the ground water once was available at the root zone of the crops now is depleting at some places. This indicator of ecological soundness was realized very well by the sample under study. The fifth rank item was "Maintenance of nutrients of soil in form of NPK and other minor elements help to have sound ecology of an area". It is observed that the farmers are acquainted with application of chemical fertilizers and count level of NPK as good indicator of ecological soundness.

Summarizing the table, it may be inferred that use of organic manures, bio-pesticides and fungicides, good soil condition, traditional farming system like crop rotation, maintenance of nutrients in soil and making ground water available at root zone of crops are the major elements to keep ecology sustainable as per understanding of the small farm operators.

ii. Economic viability

The present day living depends upon successful use of natural resources. It is argued that for development of society the natural resources have to be manipulated otherwise development cannot be sustained. To have immediate economic benefit there is tendency among people to manipulate and use natural resources without thinking of future generation. The different aspects of economic viability were examined on a five point scale as contained in the table below.

Table 53: Economic viability and reaction of the sample

Sl.No.	Statements	Average	Rank
1.	Level of production and its stability are important factors of economic feasibility that leads to sustainability	3.49	I
2.	Conservation of natural sources is an important factor to have economically feasible agriculture	3.46	II
3.	Increase in profitability in the existing farming system leads to economic feasibility	3.39	III
4.	Sustainable agriculture provides enough income to meet the requirement of family	3.36	IV
5.	Livelihood management consisting of animal and crop component on a profitable manner indicates the economic feasibility leading to sustainability	3.31	VI
6.	Cost effectiveness of technology	3.32	V
	Average	3.39	

The overall obtained score is 3.39 of all the six statements qualifying attributes of economic viability. The ranking of statements reveals that level of production and its stability (3.49) is the most important factor to look at economic viability of sustainability in agriculture. It is so because the small and marginal farm families mostly depend on farming and limitations on the basis of production level and stability over the years have concerned them more. The stability of production and productivity of coastal area of Odisha is found to be unstable due to natural calamities which require a sound technology to provide economic viability of agriculture. The second position was assigned to "Conservation of natural sources is an important factor to have economically feasible agriculture". The farmers under study carry the idea of natural resources like land, water, forest and livestock if are not maintained, the economy of middle class farm families will be greatly affected. The statement "Increase in profitability in the existing farming system leads to economic feasibility" is preferred to be in third position (3.39) because economic viability for the farmers is nothing but net profit. Profitability of technology in agriculture is the foremost consideration for which farmers have assigned third position to it as component of sustainability. The fourth position (3.36) is assigned to enough income "Sustainable agriculture provides enough income to meet the requirement of family". The "enough" is operational zed in terms to meet the family requirements under rural condition. All these factors being interrelated are with slight difference in providing relative importance of perception. The fifth (3.32) and sixth (3.31) position were assigned to the statements "Livelihood management consisting of animal and crop component on a profitable manner indicates the economic feasibility leading to sustainability" and "Cost effectiveness of technology" respectively. It is natural that to have meaningful living, crop production alone can not meet the increasing financial requirement of the farm families. Crop livestock and fisheries etc are equally important to make agriculture sustainable as inter dependent components.

iii. Social accessibility

The term social accessibility is operationalized under the study as the degree to which each and every family in the locality have accessibility to common resources. The variable 'social accessibility' was considered for the study in the matter of agriculture, and the village resources which support availability of common resources.

Table 54: Social accessibility and reaction of the sample

Sl.No.	Statements	Average	Rank
1.	Sustainability in agriculture ensures accessibility of all families to consume village resources.	3.60	I
2.	Availability and use of energy in agriculture is an important component of social accessibility leading to sustainability in agriculture	3.43	II
3.	To maintain sustainability adequate technical / technological support is necessary	3.32	V
4.	Availability of capital at local level indicates social accessibility for sustainable agriculture	3.36	III
5.	Sustainability of agriculture opens wide scope in marketing of farm produce.	3.35	IV
6.	Participatory decision making for use of common resources in village is another important outcome of sustainability	3.26	VI
	Average	3.39	

Social accessibility is operationalized in terms of accessibility of all the farm families to common village resources, use of energy in agriculture, adequacy of technological support, capital, marketing and participatory decision making (3.60) followed by use of energy in agriculture, availability of capital, technological support, market facilities. The ranking of components was found to be in relative position from locality point of view. To have sustainability in agriculture in coastal areas of Odisha accessibility and allied facilities for all and use of energy in agriculture like electricity should be made available. The other factors like credit support, marketing, technical guidance and participatory approach have been duly emphasized.

iv. Humaneness

The development ultimately aims at humaneness in the society. With change of time, the concept of development centers around humaneness consisting of honesty, value system, cordial relationship, respect for honorable living, climate of trust worthiness, compassion and social solidarity. It is expected that sustainability in agriculture depends upon these qualities of humaneness. The reaction of sample farmers towards attributes of humaneness was examined as reflected in Table 55.

Table 55: Humaneness and reaction of the sample

Sl.No.	Statements	Average	Rank
1.	Humaneness comes where respect for honesty and value system operates	3.55	I
2.	It develops cordial relationship among different communities	3.53	II
3.	Sustainability implies respect for honorable living	3.31	V
4.	Sustainability can create climate of trust worthiness	3.42	III
5.	Sustainability of agriculture creates opportunity for compassion	3.29	VI
6.	Social solidarity is outcome of the Sustainability in agriculture	3.35	IV
	Average	3.41	

The sample as a whole viewed respect, honesty and value system at the top (3.55) followed by cordial relationship (3.53), climate of trust worthiness (3.42), social solidarity (3.35), respect to honorable living (3.31) and opportunity for compassion (3.29). The people of Odisha are religious and lead a simple life. The tradition of the social system in Odisha contains all these good attributes as have been ranked above. The value system in agriculture is more rigid while honour, honesty and cordial relationship among the groups and climate of trustworthiness prevail at higher rate among the people of Odisha as have been reflected as indicator of sustainability. However, creation of compassion for others is the least preferred item.

v. Adaptability of sustainable measures

Sustainability in agriculture is considered important because to keep it productive for times to come. Technologies evolve at different parts of the country and outside are to be adopted with refinement to arrest degradation of natural resources. The present study considered progressiveness, accommodation of Government policy, technological change, and increase in market demand, consumer preference, social conflict and disaster management in the context of adaptability of technology. The reaction of the sample to these attributes is given in Table 56.

Table 56: Adaptability and reaction of the sample

Sl.No.	Statements	Average	Rank
1.	Sustainability is achieved where villages tend towards progressiveness.	3.48	I
2.	Sustainability emphasizes accommodation of Government policy and technological change.	3.37	II
3.	Sustainability makes market demand widened.	3.30	IV
4.	Consumer preference is important to have Sustainability in agriculture	3.32	III
5.	Possible conflicts arise in adoption of Sustainability methods / practices and measures	3.29	V
6.	Sustainability provide scope for disaster management	3.37	II
	Average	3.35	

The progressiveness (3.48) as first factor of adaptability followed by disaster management (3.37), acceptance of Government policy (3.37) and consumer preference (3.32) to have sustainability in agriculture. The least preferred item was found to be possible social conflict next to creation of market demand for adoption of technology. These attributes are essential as have been reported by authors in field of social research. Adoption of technology without movement of people towards progressiveness carries little value for the people of coastal belt of Odisha.

7.3. Factors and sub-factors of sustainability of agriculture

Factors and the associated sub factors of sustainability of agriculture are the deciding factors of sustainability of small farm production systems in totality. The farmers perception of these factors and sub factors in respect to the small farms production systems were mentioned as perceived by the farmers under study as presented Table 57.

Table 57: Factors and sub-factors of sustainability

Sl.No.	Factor	Sub factors
1.	Ecological soundness	1. Soil condition 2. Ground water level 3. Nutrient content 4. Crop rotation 5. Use of Organic fertilizer 6. Use of bio-pesticides
2.	Economic viability	1. Production level 2. Conservation of natural resources 3. Profitability 4. Income level 5. Livelihood system 6. Cost effectiveness
3.	Social accessibility	1. Common use of village resources 2. Use of energy in farming 3. Technical support 4. Capital availability 5. Marketing of produce 6. Participatory decision making
4.	Humanness	1. Respect for honesty 2. Cordial relationship 3. Honourable living 4. Climate of trust 5. Compassion 6. Social solidarity
5.	Adaptability	1. Progressiveness 2. Policy acceptance 3. Demand driven production 4. Consumer Preference 5. Resolution of conflict 6. Diaster Management

The above mentioned factors and sub factors are taken into consideration for assessing the gap in perception about the sustainability, correlation of factors and extent of contribution of factors towards sustainability of small farm production system as well as assessing the sustainable agriculture index.

7.4.Gap in perception

Summarizing the findings of the reacted values on five important components like ecological soundness, economic viability, social accessibility, humaneness and adaptability of recommendations the gap was calculated as mentioned in Table 58.

Table 58: Gap in perception about the sustainability in agriculture (n=200)

Sl. No	Factors of sustainability	No. of statements	Max. obtainable score	Scores obtained	Gap %
1.	Ecological soundness	6	30.00	19.96	33.47
2.	Economic viability	6	30.00	20.37	32.10
3.	Social accessibility	6	30.00	20.23	32.57
4.	Humaneness	6	30.00	20.27	32.43
5.	Adaptability	6	30.00	20.15	32.83
	Average	30.00	20.19	32.70	

All the five factors tend to have a gap up to 32.70% in average, maximum being in case of ecological soundness and minimum in economic viability. The farmers being homogenous and the factors of sustainability being familiar to majority of respondents the gap appears to be same in almost all the cases. In other words about 67.00% of the samples are aware of factors like, need to have sustainability in agriculture.

7.5. Correlation of sustainability factors

The sustainability factors like ecological soundness, economic viability, social accessibility, humaneness and adaptability are found to have relationship on overall perception of sample under study. As has been indicated earlier, the samples are categorized into high, medium and low level. The possible relationship between factors of sustainability and perception was worked out as reflected in Table 59.

Table 59: Factors and perception of sustainability in agriculture (n=200) ('r' Value)

Sl. No	Factors of sustainability	Low	Medium	High	Total
1.	Ecological soundness	0.393	0.083	0.189	0.541*
2.	Economic viability	0.055	0.182	0.375	0.496*
3.	Social accessibility	0.265	0.369	0.404	0.507*
4.	Humaneness	0.139	0.089	0.612	0.553*
5.	Adaptability	0.077	0.336	0.459	0.428*

* Significant at 5% level of probability

The perception score at high level is found to be positively associated with economic viability, social accessibility, humaneness and adaptability whereas ecological soundness was found to be not significant.

In case of middle level perception social accessibility and adaptability were found to be co-related while at lower level it is only ecological soundness and social accessibility. In other words, ecological soundness at high level and medium level is not found to be positively associated with perception as in case of low level. However, at overall level all the factors exhibited positive correlation indicating that all the five factors as described above have direct relationship with formation of perception about sustainability in agriculture by small farm operators.

Table 60: Percentage contribution of factors of sustainability towards Sustainable agriculture

Sl. No.	Factors of sustainability	'r' – Value	R^2 x 100
1.	Ecological soundness	0.0541	29.22
2.	Economic viability	0.496	24.59
3.	Social accessibility	0.507	25.67
4.	Humaneness	0.553	30.58
5.	Adaptability	0.428	18.31

The contribution of factors of sustainability in agriculture with overall perception level is observed as depicted in Table 60 and Figure 5.

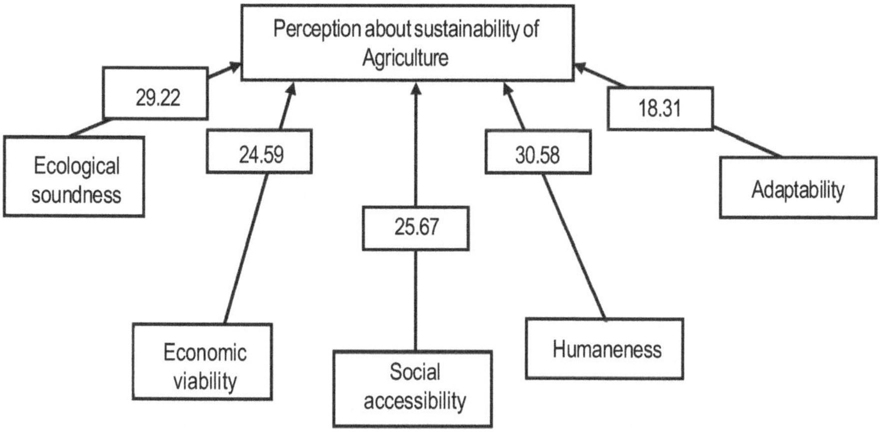

Fig. 5: Perception about Sustainability of Agriculture

In forming perception, the factors like humaneness contributes maximum (37.46%) followed by adaptability (21.03%), social accessibility (16.32%), economic viability (14.43%). However, the contribution of ecological soundness was observed to be minimum (3.57%). The reasons for such findings account for (i) humaneness in terms of living, (ii) adoption of technology for progressiveness, (iii) social accessibility for use of common resources and (iv) profitability. All these four factors are directly or indirectly concerned with human beings in living system. Therefore, while forming perception, the humaneness factor surpassed the factor of ecological soundness.

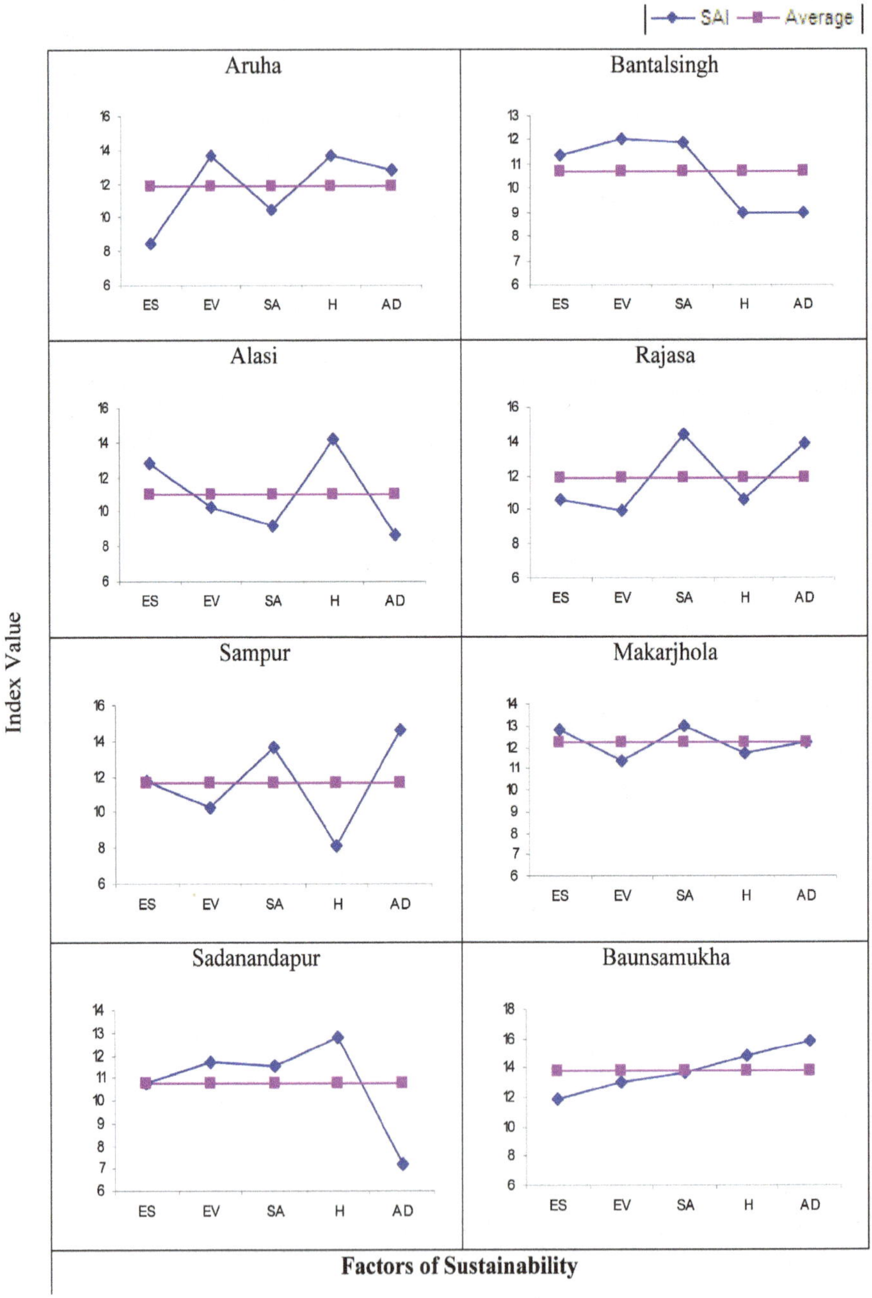

Fig. 6: Sustainable Agriculture Index of Villages under Study

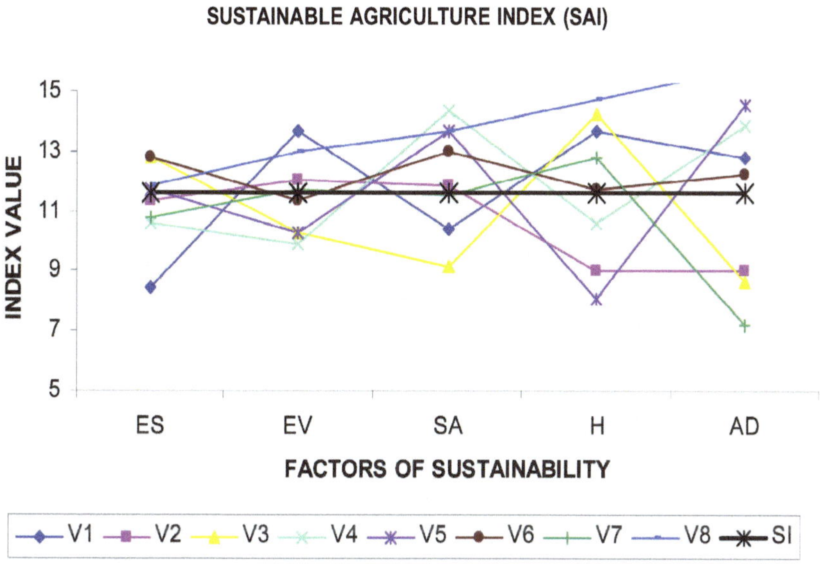

V1 – Aruha	V2 - Bantalsingh	V3 - Alasi
V4 - Rajasa	V5 - Sampur	V6 - Makarjhola
V7 - Sadanandapur	V8 - Baunsamukha	SI – Sustainable Index

Fig. 7: Sustainable Agriculture Index (SAI)

8

Adoption of Recommended Measures of Sustainable Agriculture

8.1. Ecological soundness

Conservation of natural resources essentially means maintenance and improvement of ecological status and not permitting them for further degradation. The natural resources are soil, water, forest, mountain and animal resources. An attempt was made to find out whether the respondents adopt certain recommendations for preservation of these natural resources irrespective of source of advice.

i. Soil health

Soil is considered as an important ingredient of ecological soundness of an area. Adoption of soil health measures, like adding of organic manures, following of crop rotations, use of balanced chemical fertilizers, green manure and bio-fertilizers are some of the recommendations to maintain soil health. While examining the adoption of these measures, the following results were obtained.

Table 61: Adoption of soil health measures (n=200)

Sl. No	Soil health measures	Adoption	
		Frequency	Percentage
1.	Use of organic manures like compost, FYM etc.	61	30.50
2.	Crop rotation	130	65.00
3.	Balance use of chemical fertilizer	0	0
4.	Green manuring	80	40.00
5.	Bio-fertilizers	81	40.50

The percentage analysis of data presented revealed that the sample under study are adopting crop rotation (65%) followed by use of bio-fertilisers (40.5%), green manuring (40%) and organic manures to retain health of their soil. However, it is surprising that none of the farmers test soil and then apply chemical fertilisers in balanced dose. The use of organic matter in soil particularly was observed to be age old practice with the sample.

ii. Water

Odisha suffer from erratic rainfall almost every year. The recommendations are given to the farmers to preserve run-off water and to make optimum use of water to raise the crops. Adoption of recommendations with regard to water use technology was examined as given in the Table 62.

Table 62: Adoption of water conservation measures

Sl. No	Ingredients of Water conservation measures	Adoption	
		Frequency	Percentage
1.	Check run-off of water through small and medium water reservoirs	72	36.00
2.	Raising of vegetations	160	80.00
3.	Creation of irrigation facilities	143	71.50
4.	Water harvesting structures (in situ water management)	73	36.50

Farmers in the area of study are much more acquainted with growing of vegetation (80%) followed by creation of small irrigation facility (71.5%) and checking run-off of water (36%). However, water harvesting structures is found to be adopted by 36.5% of the sample. The farmers are conscious about preservation of rain water in the small water reservoirs, like tanks and ponds. From time immemorial, some tanks and small water reservoirs were created to arrest run-off water which are in bad conditions.

iii. Bio-control methods

In recent years there has been consciousness among the farmer to take up bio-control methods for preserving ecological soundness. It was observed that out of 100 small farmers 57.5% have adopted bio-pesticides and bio-fertilisers to replace chemical pesticides which add preserve to ecological entity of the area.

Adoption of conservation measures in respect of soil, water, forest, mountain and animal are summarized in Table 63.

Further, making comparison of adoption of conservation methods between districts, the results revealed that preservation of animal is found to be maximum followed by soil and water. This is so because; the introduction of cross breed or up-gradation of local breed has not been taken up extensively. People have preference for local breeds especially in case of cows for the reasons of maintenance, free from diseases and feed consumption. However, absence of forest and mountain in the area under respective districts did not intend the farmers to adopt measures covering these. However, the statistical information as per District statistics handbook, Govt. of Odisha (2003) reveals that forest area of the districts of Balasore, Khurda, Puri and Ganjam is reported to be 0.57%, 1.064%, 0.235% and 5.41% respectively.

8.2. Economic viability

The economic viability is the composite outcome of different measures adopted by the farmers to increase the yield and thereby profit. The following table reveals the adoption of productive ingredients in the area under study to increase profitability of farming.

A look at the table reveals that use of HYV seed is found to be extensively used by respondents under study, where as irrigation management practices as recommended are adopted by 60%, post harvest technology by 39.5% and bio-fertilizer by 40.50% of the sample. However, no response was recorded with regard to use of balanced chemical fertilizer and adoption of INM and IPM practices. The practices of INM and IPM have yet to be popular with the farmers.

Table 63: Adoption of measures for preservation of natural resources

Sl.No	Elements of Natural resource	D1		D2		D3		D4		Total	
		f	%	*F*	%	*f*	%	*f*	%	*f*	%
1.	Soil	18	36.00	19	38.00	31	62.00	22	44.00	90	45.00
2.	Water	19	38.00	18	36.00	24	48.00	22	44.00	83	41.50
3.	Forest & Mountain	—	—	—	—	38	76.00	—	—	38	19.00
4.	Animal wealth	25	50.00	22	44.00	25	50.00	25	50.00	111	55.50

DL : Puri D2 : Khurda D3: Ganjam D4: Balasore

Table 64: Difference among the districts in the preservation of natural resources

Sl. No.	Elements	D1 X D2	D1 X D3	D1 X D4	D2 X D3	D2 X D4	D3 X D4
1.	Soil	0.207	2.601*	0.816	2.40*	0.610	1.803
2.	Water	0.207	1.01	0.610	1.22	0.816	0.803
3.	Forest & Mountain	—					
4.	Animal wealth	0.601	2.92*	0	3.485*	0.601	2.92*

* Significant at 0.05 level
Where, D1 : Puri D2 : Khurda D3 : Ganjam D4 : Balasore

Table 65: Adoption of productivity components

Sl. No	Production Ingredients	Adoption	
		Frequency	Percentage
1.	HYV Seeds	200	100.00
2.	Balance use of fertilizers	0	0
3.	Irrigation management practices	120	60.00
4.	Post harvest technology	79	39.50
5.	Use of low cost inputs like bio-fertilisers/ organic manures to replace chemical fertilisers	61	30.50
6.	INM and IPM practices	0	0

i. Profitable crop-husbandry

Sustainability in agriculture ultimately aims at profitability of present farming systems without endangering the future. In finding out the measures adopted to increase the profitability of crop-husbandry the following results were obtained.

Table 66: Adoption of profitable measures in Crop-husbandry

Sl. No	Components of profitability	Adoption	
		Frequency	Percentage
1.	Adoption of package of practices	89	44.5
2.	Use of latest information in farming	99	49.5
3.	Contact with marketing network	112	51.00
4.	Value addition	85	42.5
5.	Emphasis for consumer- preference	75	37.5

There is increasing trend among the farmers to establish contact with the market and make use of latest information in farming followed by adoption of improved package of practices and making value addition to fetch more price in market. However, 37.5% of sample found to have interest to grow crops keeping consumer preference in view. The trend reveals that farmers are gradually becoming interested in commercializing farm produce in the area under study.

ii. Profitable animal husbandry

Livestock continued to be supporting source of income for the farm families in rural areas of Odisha. With rising of consciousness for better living, a diversification in occupation is observed. The livestock is found to be most acceptable enterprise for the farmers even at micro level to increase income level. The information collected regarding efforts made by sample farmers to increase the profitability in animal husbandry following information was received.

Table 67: Adoption of recommendations leading to profitability of animal husbandry

Sl. No	Components of profitability in animal husbandry	Adoption	
		Frequency	Percentage
1.	Quality breed	109	54.50
2.	Use of balance feeds	55	27.50
3.	Raising of fodder	48	24.00
4.	Health care practices	48	24.00
5.	Use of technical information	47	23.50
6.	Value addition	76	38.00
7.	Market link	79	39.50

Analysis contained in table above reflects that the farmer in area under study are relatively more conscious in adopting quality breed (54.5%), establishing linkage with market due to operation of OMFED (39.5%) and value addition (38.0%) followed by use of fodder and health care practices (24.0%). Use of technical information was found to the extent of 23.5%. The study shows that components leading to profitability to animal husbandry have been adopted by the sample to a level of satisfaction.

8.3. Social accessibility

Social accessibility is another aspect of sustainability in agriculture. In examining the aspect of study in terms of equal access to common resources, use of energy, provision of adequate technological support, availability of capital to operate the occupation and prevalence of participatory decision making for common use, the following results have been obtained.

Table 68: Elements of social accessibility

SL. No		Present		Not present	
		Frequency	%	Frequency	%
1.	Equal access to common resources	101	50.50	99	49.50
2.	Use of energy in farming	109	54.50	91	45.50
3.	Technological support to all	100	50.00	100	50.00
4.	Easy availability of credit to all	97	48.50	103	51.50
5.	Community decision to adopt new technology	96	48.00	104	52.00
6.	Equal market opportunity to all	100	50.00	100	50.00

Results reveal that the prevailing situation in case of community decision making, easy availability of capital in farming, equal marketing facility and easy access to common resources in the villages are at discouraging rate.

8.4. Humaneness

Human value and honorable living are interactive factors in deciding worth of living in society. Sustainability in agriculture aims at providing these human values to the practitioners. While examining the humaneness in the term of respect for honesty, good relationship, respect for honorable living, climate for trustworthiness, feeling for others and social solidarity, the following results were obtained.

Table 69: Elements of humaneness

SL. No	Elements of Humaneness	Present		Not present	
		Frequency	%	Frequency	%
1.	Respect for honesty in society	89	44.50	111	55.50
2.	Good relationship among the communities	88	44.00	112	56.00
3.	Respect for honorable living	87	43.50	113	56.50
4.	Climate of trust worthiness	94	47.00	106	53.00
5.	Feelings for others	78	39.00	122	61.00
6.	Social solidarity	95	47.50	105	52.50

The elements of humaneness, which constitute our social value system are not in desirable manner. Almost all the elements are found less than 50% than the expectations. Feeling for others, respect for honorable living and cordial relationship among the communities need a good deal of effort to add to sustainability. These indirect factors have much more influence on sustainable agriculture index (SAI) because agriculture is an enterprise of human values in our society.

8.5. Adaptability

Adaptability is a measure of accommodation assimilation and integration of new technology into social system. Proneness to accommodate changes in agriculture, Govt. policy, changing market situation, consumer preference, conflict to adopt and technique of disaster management were studied as revealed in Table 70.

Table 70: Adaptability of sustainable components in farming

SL. No	Elements of Adaptability	Present		Not present	
		Frequency	%	Frequency	%
1.	Technological change in agriculture	106	53.00	94	47.00
2.	Market demand	103	51.50	97	48.50
3.	Consumers' preference	99	49.50	101	50.50
4.	Conflicts in adoption of sustainable measures	79	39.50	121	60.50
5.	Disaster management	59	29.50	141	70.50

Analysis of the table infers that Govt. policy and technological change in agriculture are accepted by the people for making quality livelihood. Adaptability to the changes in market demand, preference to consumer and avoiding conflict in adoption are found to be in order indicating the reaction of the people. In sustainability providing scope for disaster management was responded by 29.5% of the sample. On the whole the table depicted that there is increasing trend among the people to accommodate adaptability of sustainability measures.

Summarizing the findings on adoption of different sub-components of sustainability of agriculture, the following results were obtained.

Adoption and preference of sub-components of sustainability in agriculture

The adoption and preference of farmers' in respect of sub component of sustainability of agriculture were summarized and presented as follows.

Table 71: Adoption and preference for sub-components of sustainability in Agriculture

Sl. No.	Sub-components of Sustainable Agriculture	Average (%)
1.	Ecological soundness	45.60
2.	Economic viability	41.70
3.	Profitable crop husbandry	41.33
4.	Profitable animal husbandry	33.00
5.	Social accessibility	50.25
6.	Humaneness	44.33
7.	Adaptability	46.17

Adoption of social accessibility, adaptability, ecological soundness, humaneness, economic viability, profitability of crop husbandry and profitable animal husbandry are found to be in order.

9

Agribusiness Opportunities for Sustainable Farm Production

At present producers and consumers are concerned about products and their uses. In farm sector agribusiness is gaining momentum because of marketing of farm produce and its availability to the consumers at affordable rate. Agribusiness thus has been identified as important element in improving farm business. The farm business is otherwise known as agribusiness. Agribusiness is an off farm activity mainly concerned with processing and marketing. It implies that in addition to production of farm produce farmers responsibility increases to process the out-put and its sale in the market for better price. The farmers like to approach different organizations, corporation and regulated markets. Therefore attempt has been made at government level to encourage cultivators to invest in the farm business. In order to increase the efficiency of farmers there is need of keeping the business flexible encouraging farmers to invest in farm and allied sector business.

9.1. Agribusiness inputs

Supply of inputs like fertilizers, plant protection materials etc. In addition agribusiness focuses on use of organic matter in farming like compost, oil cake and other bi-products. To encourage and sustain their activities, the Government is creating appropriate infrastructure facility. In short agribusiness is a term which infers supply of farm inputs and sale of farm outputs to provide benefits to the small farm operators who form the bulk of the farming community.

Agriculture, horticulture, fishery and livestock come within the actual domain of agribusiness.

The concept of agribusiness has been referred from management point of view where different types of management are required to play their role. These are; poduction management financial management, marketing management, personnel management, material management, purchase management, whole sale management, farm management, office management and export management etc. All these management processes have applicability in business management in general and agribusiness management in particular for successful business in farm sector. The economic resource and system of authority are two important aspects in management of farm and its products. The management features like decision making, organizing, staffing, planning, controlling, communication and direction are essential parts of agribusiness.

To run the agribusiness a common farmer needs skill and competency. Therefore the desirable traits of good agribusiness manager are knowledge, decision making power, self reliance, self assertion, social sensitivity and emotional stability. All these qualities, characteristics, nature and status together explain the concept of agribusiness management.

9.2. Key factors of small farm operation and sustainability

Agribusiness is a composite unit of number of factors. The key factors are many and multidimensional in planning the farm activities particularly at small farm level it becomes necessary. In addition the manager of the farm should know the potential of his farm in production productivity and cost recovery. Accordingly, the farm should be managed for meeting the demand of the market. Planning for small farm operation and emphasis on sustainability are two important factors for long term profitability. The farmers in are general are expected to be aware of such key factors for successful farming. In planning farm activities it becomes necessary for the status of farm in terms of (i) human aspects, (ii) Physical aspects and (iii) financial aspects.

i. Human aspects of agribusiness

Human aspects of agribusiness in family based and developed in line of the family, status, culture and aspiration where the main aspects of farm management need to understand the following aspects.

- Farm family goal
- Skill of the farm family
- Family labour

- Family linked to development organization
- Proximity of the family to the market.

Farm family goal

Goal of the family, objectives and aspiration are linked together to express status index of family. The scope of change in farming, farming satisfaction, business opportunity is required for maximization of profit. In agribusiness management, human factors are to be considered in which market survey to be determined through participatory approaches. The farm family needs change. Therefore direction of change and utility of change requires deep analysis before planning for the farm.

Skill of the farm family

At present farming is a skilful job. It is taken as skilful activity after green revolution in making balance use of fertilizer to obtain desired output. The extension principle "where people are and what they have" is the key to plan farm activities on the basis of skill and management efficiency.

Farm family labour

Labour is major component in management of farmer. Whatever may be the use of equipments and tools of farming, the labour requirement will continue for a very long time under Indian condition that to more in condition of coastal ecosystem in Odisha. Better utilization of farm labour, diversification of farm activities and non-farm activities are considered in a skilful manner keeping agro climatic conditions in view in order to increase productivity potential and profit.

ii. Physical aspects of agribusiness

Physical aspects include the following factors.

1. Cropping
2. Maintenance of soil fertility
3. Cropping labour needs
4. Present utilization of market
5. Irrigation
6. Livestock
7. Farm power

Cropping

Cropping pattern of a farm is season specific and use term as kharif, rabi and summer. The cropping system in cropping intensity like sole crop, double crop, crop rotation, intercropping, mixed cropping, relay cropping and multiple cropping are the basketful of choices depending upon agro climatic zone and other essential factors. There are many bottlenecks which may hinder utilization of the potential. These are mainly inadequate and irregular rainfall, labour shortage, lack of technological support and perpetual disease and pest. However in small farm operation system, the affordability of farmers and external credit facilities are considered to prepare farm planning.

Maintenance of soil fertility

Soil is the soul of farming. Fertility status of soil is of prime importance in sustainable farming. The methods followed by farmers are legume and non legume based crop rotations, use of manures in cultivation, turning of soil by mould board (MB) plough, use of chemical fertilizers and control of weeds through mechanical and chemical means. The recommendation is to increase productivity and production percentage of different crop rotations. Maintenance of soil fertility through natural means and use of longer and better fallow system are recommended. But keeping the land fallow in case of small and marginal farmers seem to be difficult, for the land is the major source of farming for them. The Provision of organic manure from outside to their soil is another method to sustain soil fertility. The recommendation is that chemical control of diseases and pests should be minimized and encouraging plant protection by organic means. The constraints are there which are of location specific. The seeds of improved varieties are used to get more output, for instance improved variety of sugar cane sets are now coming to market. The potential, is that planting materials are available in sufficient quantities that create constraints to opt for organic farming.

Cropping labour needs

Indian farming is depending human labour. In spite of attempt for farm mechanization the labour needs remain to continue in terms of number. The labour may be skilled or unskilled. With increase in labour cost the farm and farmers are faced with financial problems. There are methods by which the demand for labour can be reduced for maximizing the profit and minimizing the cost of production. This requires proper crop planning. rotation and seasonal farm activity.

Crop product utilization and marketing

The marketing channels usually existing in India are farmer-village or itinerant trader and whole saler-retailers–consumer. On small farm a high portion of the product is domestically consumed or in case of sales it may be a distressed sale resulting in buying back at a higher cost. In case of large farmers a large portion goes to market. On the farm usually processing is not very much prevalent but in some cases it is done for instance, milk is processed into butter fat (*ghee*), dahi and khoa, paneer etc. In case of rice (processed paddy) and flour from wheat, gram, pea are prepared for domestic consumption rather than sale. Storage is done in a very small scale in mud bins, drums, gunny bags etc., and in general there is high percentage of loss due unscientific storage. Potential lies by way of improving the presentation by resorting, grading and cleaning the food grains before they are sent to markets. The farmers need to know to market their products through marketing societies or regulated market. There is huge scope for improvement in marketing. The loss in storage can be reduced to a greater extent by adopting the scientific storage by building moisture, rodent, and insect-pest proof infrastructure individually / for community. The farmers need to restore cooperative marketing to prevent them being exploited by middlemen much dominant at present local marketing system.

Irrigation

Major sources of irrigation in India are: rivers, canals, tanks and water reservoirs, tube wells, masonry wells and artisan wells (in hilly areas). At present the major sources much utilized are canals, tube wells and pumping sets. The statistics of irrigation shows that the potential of irrigation is much more than the utilization. The increase in development of canals has brought a definite change in cropping pattern. Presently in order to cope with the scarcity of water, recycling of water is practiced. In the areas with shortage of irrigation water, there is use of sprinkler and drip irrigation system to use water for irrigation in a judicious manner.

Feed supply for livestock

Stall – feeding is the usual and most popular method. The sources are pasture or grazing in the open. However the pasture does not have proper vegetation to full fill the needs of the livestock. There is a lot of potential for cultivation of fodder crops and planting of pasture lands although there is a keen competition between crop and animals for the land having a short supply in India. The farmer running a mixed farm should demarcate the land into fodder production and food grain production and cultivate high yielding fodder crops for economizing on land, use of fertilizers and other organic matter for better output of fodder crops. Now a days there are certain fodder trees which could be planted on waste lands, on the field boundaries, in front of houses and fodder is obtained to feed the livestock. Balance diet is very important to economize the fodder crops.

Farm grown and purchased feed from factories

The feed which is mixed in the farm comes from the farm or from the market. The balance ration is the best ration. The dairy farmers though are having knowledge of the balance ration, but they hardly calculate the proportion of feed ingredients while preparing feed mix. Similarly the scientific way of preparing ration in home is required to get better performance. There is always scope of improving the quality of fodder by following the advice given by extension agent and referring farm literatures on livestock production. The farmers should take the advantage of television and radio programmes organized for farmers in this field. Extension agents required to work sincerely in feeding the livestock farmers on latest development in livestock production and management.

Animal activities

Much has been done improving breeds of livestock especially in the case of milch cattle and buffaloes. Hybrid chickens are available for broiler industry. Some improvement was done in draught animals. Researches on livestock production and management have brought a lot of beneficiaries in the field of animal husbandry and dairying. Much attempt has been made to control the diseases particularly in epidemic form specifically in poultry and they are now safe from prevalence of epidemic. In cattles the diseases are controlled by vaccinating in time. However the diseases and pests are of great risk in livestock business. The management which includes feeding, breeding, health and hygiene is very important. In case of feeding balanced ration brings a lot of improvement. Meat animals and broiles also need balanced ration for appropriate marketable growth. In this cross breeding is gaining importance and artificial insemination is emphasized for improving the milk output.

Options of small farm operators towards agribusiness under the study

The sample under study (n=200) were interviewed to reveal their option in terms of (1) Human aspects, (2) Physical aspects and (3) Financial aspects for a sustainable small farm agribusiness.

iii. Financial aspects of agribusiness

The technological development has increased the aspects of large margin of profit. In the small and marginal farms whatever capital investment is in the land and bullock, plough, some small implements, kaccha cattle shed and a cow or buffalo in comparision to larger farms investment in lend (fixed) inherited, pucca cattle shed, barn, implement house, tractors, ploughs and other tillage implements, tube wells, pumping sets, storage, implement shed etc. with well regulated output disposal. The small and marginal farmers do not have enough marketable surpluses as in case the large farmers who sell their products through marketing channels. The small farm operators can take the benefit through cooperative efforts in order to run agribusinesses. The farmers if use their resources judiciously, the cost of production would be minimised and profit maximisation will take place. They may take the advantages of regulated markets through cooperatives as it is happening in the states of Maharastra and UP. The case of small and marginal farmers however is to be considered in front of agribusiness venture and financial management to ensure sustainability of farm business in terms of profitability and economic viability for which a proper viable farm plan must be followed for the entire year.

9.3. Human aspects for farm development

The human factor is of first importance as he is the manager of the farm. The skill and attitude lead to success or failure in farm business. The factors are related to farmer himself in relation to environment in which he resides and work. As evidenced from the findings presented in the following table that there was as high as 66% gap in family labour use for farm management which indicated that the small farm operators need to be dependent on outside labour for their farm business management. This external dependency may affect the profitability of the farm business.

Table 72: Human factors and gap

Sl. No	Factors	Mean Score	GAP (%)
1.	Family goal	2.15	28.33
2.	Skill level	2.05	31.66
3.	Family labour use	2.03	65.66
4.	Educational level	2.10	30.00
5.	Interest of family members in farming	1.58	47.33
6.	Social status	1.90	36.66
7.	Favourable farming climate	2.08	30.66
8.	Awareness about recent technologies	1.83	39.00
9.	Information sources	1.78	40.66
10.	Availability of technical support	1.75	48.33

Farmer's option also showed a gap of 48.33 % and 47.33 % in availability of technical support and interest of family members in farming respectively. The technical support as well as the interest of family members are considered to be the driving factors for business activities. Hence these factors having a higher gap should be reduced for sustainable agribusiness of small farms. Awareness about recent technologies and the social status of the farmers also revealed gap of 39 % and 36 % respectively indicating that the farm business management require interest of farm family in agribusiness venture. The social status of small farm operators may not be the sole deciding factors as because the farm business is more an entrepreneurial action than cultivation only indicating a varied social status.

The small farm operators also posed gap in skill level (31.66 %) and educational level (30.00 %). The gap is more or less 30 % which revealed that there is scope for improvement for business activity through proper training and exposure in the field of suitable farm business development and management. Gap in favourable farming climate was to the tune of 30.66 % as revealed from the study. This is not so high and also keeps opportunity of managing with location specific technologies. Hence the climate suitability is to be examined and proper measures may be taken addressing the climatic uncertainty to ensure a protection against any climatic aberrations for farm business activities.

9.4. Physical aspects of farm development

Physical aspects includes the important factors like crop, cropping pattern maintenance of soil fertility and irrigation. The small farm operators have placed their view points as presented in Table 73.

Majority of farmers accepted both short medium duration and so also the long duration crops as per the availability of irrigation facility. How ever the farmers have adopted sole crops looking at the scanty irrigation water supply during kharif season as implied from their adoption pattern of crops and cropping pattern and irrigation supply to the crops grown. How ever all most all the farmers have their farm partially irrigated irrespective of source.

Animal Husbandry

Animal husbandry is one of the most important aspects in farm business sector. All most small the small and marginal farm families possess livestock; but care taken for animals is lacking so far commercial livestock farming is concerned which is high in case of the larger farmers compared to the case of small. The present study attempted to gather information on livestock management practices followed by the small farm operators as mentioned in Table 74.

Table 73: Physical aspects of farm development

Sl.No	Physical factors	Level of adoption of recommended measures			Mean Score
		Full	Partial	Not at all	
1. Crop					
I	Short duration	120	60	20	2.50
Ii	Medium duration	80	40	40	1.80
iii.	Long duration	75	I25	15	2.45
2. Cropping pattern					
I	Sole crop	120	30	50	2.35
Ii	Double crop	80	40	30	1.75
iii	Crop rotation	100	20	80	2.20
iv	Inter cropping	10	80	110	1.50
v	Mixed cropping	19	18	173	1.38
3. Maintenance of soil fertility					
I	Legume based rotation	100	76	24	2.38
ii	Use of organic matter	128	65	7	2.64
Iii	Use of chemical fertiliser	00	200	00	2.0
iv	Use of crop residue	32	168	00	2.16
4. Irrigation					
i	All farm	25	175	00	2.0
Ii	All crops	25	175	00	2.0
Iii	All seasons	23	177	00	2.0

The findings revealed that majority of the farmers did not adopt the critical livestock management practices i.e. regular check up and treatment, balance feed, clear drinking water, cattle shed management and preparation of milk and milk product for family consumption and sale. That might be reducing the productivity of the animal wealth. Hence the farmers required to be trained and supplied with the critical inputs to exploit this opportunity of livestock rearing in mode of sustainable agribusiness in situ for better income and profit.

Table 74: Animal husbandry

Sl.No	Factors	Adoption	Non- Adoption
1.	Regular check up and treatment	40	160
2.	Balance feed	20	180
3.	Clear water for drinking	30	170
4.	Cattle Shed management	15	185
5.	Preparation of milk and milk product	25	175

Possession and use of farm machineries

Few of the farmers under study were possessing advanced farm equipments; tractor, power tiller, power weeder, power sprayer, and winnower / thresher were possessed by 6, 18, 11, 22 and 34 numbers farmers respectively. However none of the small farm operators under study possessed any one of harvester / reaper and sprinkler. This trend put forth the promotion of use of farm implements appropriate for the farming situation for time and cost saving efficiency in farming which in turn create opportunity of developing agribusiness activities as they get surplus time for operating such economic activities.

Table 75: Possesion and use of farm machineries

Sl.No	Factors	Possessors	Non-Possessors
1.	Tractor	6	194
2.	Power tiller	18	I82
3.	Power sprayer	11	191
4.	Power weeder	22	178
5.	Harvester / reaper / sprinkler	0	200
6.	Winnower / thresher	34	166

9.5. Financial aspects for farm development

The financial aspects are the key to farm business development and management. The financial sources for agribusiness required to be accessible by the small farm operators. One should avail the opportunities of finance offered by different developmental schemes of Government. The farmers may meet the investment from their own source or from credit or from both the sources. However the small farm operators for the farm business may face difficulties in investing from their own and rather go for investing from credit with relaxation of interest and other incentives offered by the Govt. The information was elicited from small farm operators under the study in respect of their option on financial aspects towards agribusiness ventures. The findings evidenced from the Table 76 indicated that only five farmers opted for use of own source of funding fully for running farm business activities.

Table 76: Financial aspects

Sl.No	Financial sources	Full	Partial	Not at all
1.	Own resource	5	25	170
2.	Credit	160	20	20
3.	Own resource and credit	190	10	0

Although majority (170) small farm operators opted not to use own source of finance at all and twenty farmers opted use credit at all; the majority of small farm operators opted meeting financial requirement from both own source and credit to run agribusiness as a sustainable venture. The agribusiness activities hence need to be promoted both through technological and financial support (own source and credit based). The human resource development in this context keeps priority.

10

Extension Approach and Sustainability in Small Farm Production System

Transfer of technology is closely associated with generation of technology and its use leading to development. Extension components contain stimulating inputs that would mobilize the practitioners to put recommendations into action. Sustainability in agriculture is very important component of present day agriculture and future aspirations. Keeping these aspects in view an attempt was made to enlist suggestions under extension approaches that would enhance sustainability in agriculture in the area under study. The suggestive extension measures to enhance sustainability in agriculture were studied under components of social structure, cooperation mechanism, evaluation technique and marketing network.

10.1. Social structure and organization

To initiate action in the social system there is need to intervene the social structure and a structural set up is to be devised to put recommendations into action. Sustainability in agriculture needs community approach and initiative. While examining the possible social structure either existing or can be developed to maintain Sustainability in agriculture the following results were obtained.

Table 77: Suggestive social structure in order of importance

Sl. No.	Dimensions	Rank
1.	Committee with specific regulations	I
2.	Local organizations and NGO	II
3.	Government's line departments	III
4.	Youth association	IV
5.	Women forum	V
6.	Agri-business / traders	VI
7.	Agri-clinic	VII
8.	Local credit institutions	VIII
9.	Involvement of ATMA of State department of Agriculture	IX
10.	Others interested in preservation of eco-system	X

ATMA: Agricultural Technology Management Agency

These ten important social structures can be utilized to infuse idea of sustainability of agriculture among the small farmers and insist upon for follow up.

10.2. Cooperation mechanism

In rural Odisha there are many organizations and institutions which start operating but closed down after sometimes owing to reasons of many factors like lack of leadership, lack of resources and absence of remunerative development programmes. Attempt was made to find out as to what kind of cooperation mechanism is necessary to sustain the sustainability of agriculture.

Table 78: Cooperation mechanism for sustainability of agriculture in order of importance

Sl. No.	Dimensions	Rank
1.	Regular meeting of sustainable committee	I
2.	Inviting experts for technical support	II
3.	Organisation of training on sustainability of agriculture	III
4.	Field visit of developed area of sustainable agriculture	IV
5.	Fixing achievable objectives	V
6.	Developing a plan of action	VI
7.	Incorporation of Govt. Policy for sustainable agriculture	VII
8.	Contact with outside agency for suggestion	VIII
9.	To establish linkage with development department	IX
10.	Encouraging farmers to offer innovative ideas	X

Suggested mechanisms(Table 78) seem to be logical and sound as it involves local people to the maximum extent to work towards sustainability of agriculture. Inviting experts from outside, training, field trip, linkage with development departments and working on innovative ideas of farmers can show better path on how to maintain sustainability. However, establishment of achievable objective and developing sound action plan will bring success to preserve eco-system.

10.3. Evaluation technique

Evaluation is an integrated and important component of any programme to be undertaken. The community approach to have sustainability in agriculture through social structure and mechanism need evaluation to indicate movement in right or wrong direction. The evaluation of action for sustainability in agriculture as suggested by the sample is recorded in Table 79.

Table 79: Evaluation technique for sustainability of agriculture

Sl. No.	Dimensions	Rank
1.	Periodical visit of experts and their opinion	I
2.	Analysis of situation	II
3.	Evaluating level of production and productivity	III
4.	Assessment of changes in socio-economic status (improvement in socio economic status	IV
5.	Assessment of progress as per plan of work at regular interval	V
6.	Modification in program on the basis of feed back	VI
7.	Documentation of reasons for failure and success	VII
8.	Re-fixing of priority	VIII
9.	Assessment of extent of people's participation	IX
10.	Study of reaction of people in general	X

The above suggestions (in order of merit) are logical and appropriate so far in the area under study is concerned. The periodical visit of experts and offering opinion, level of increase in production and productivity and status of living (improvement in socio economic status) are the real indicators of sustainability in agriculture as these are related to ecological soundness, economic viability, social accessibility, humaneness and adaptability. The above evaluating process if followed regularly the authors feel the sustainability can be maintained in the field of agriculture.

10.4. Marketing network

At present day, business in agriculture can hardly be thought of without market network, as has been reported in various studies that marketing is positively correlated with production level. The study realizing the importance of market, suggestions were sought for networking of marketing at micro-level.

Table 80: Marketing network for sustainability of agriculture in order of importance

Sl. No.	Dimensions	Rank
1.	Price fixation by local committee	I
2.	Opening of Krushak bazaar	II
3.	Purchase of farm produce by Govt. agencies	III
4.	Facility for storage and disposal	IV
5.	Price on the basis of value addition	V
6.	Display of price rate of different commodities	VI
7.	Change in fixing price policy	VII
8.	Creating facility for grading and packaging	VIII
9.	Use of market information	IX
10.	Initiation of regulatory market system	X

The above suggestions were received to make marketing network stronger to help the small farm operators to strengthen their interest level leading to sustainability.

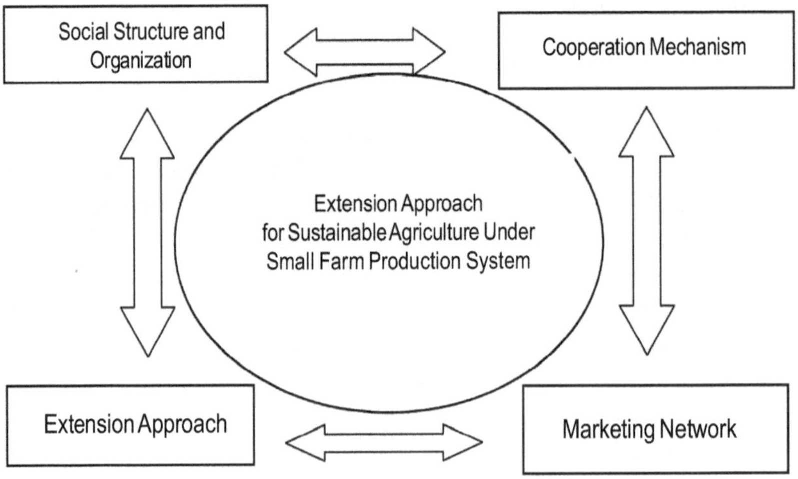

Extension Approach for Sustainable Agriculture Under Small Farm Production System

11

Strategy for Sustainable Agriculture in Small Farm Production System

The factors affecting sustainability of agriculture under small farm production system in coastal ecosystem of Odisha was studied to ascertain the status of ecological factors and suggest means and ways to improve the situation. The response of the sample farmers and the findings of the study together led to suggest the following measures for up-gradation of ecosystem of the coastal area leading to sustainability of agriculture. Like other states, Odisha is also facing with problems of sustainability of agriculture. Our sample districts namely Puri, Khurda, Ganjam and Balasore which are situated in coastal eco-system of the state has different problems concerning to natural resources like soil, water, forest and animal wealth, etc.

The suggestions to improve the sustainability status of agriculture have been viewed from two important angles. One is adoption of technological measures to preserve natural resources and other is the policy and extension approach to educate people about the necessity of taking such measures for sustainability of agriculture and improve upon.

11.1. Adoption of technological measures

Technological measures concerning to soil, water, forest and animal wealth have been advocated by the concerned departments of the state and national Govt. which have equal applicability under specific agro-climatic zone.

i. Soil status

The districts under study namely Puri, Khurda, Ganjam and Balasore cover about 11.76% of total geographical area of the state. The cultivable area of these four districts is 96,05,574 ha. The cropping intensity of these districts is 163.5% and the cropping pattern is mainly rice-pulse, rice-oilseeds, rice-vegetables and rice-fallow. The soils of these districts are not responding to as previously without use of chemical fertilizers and even to some extent over the recommended doses.

The farmers are in view that following treatments may be taken up to upkeep the soil health and with gradual removal of chemical fertilizers. The suggestions to improve soil condition are given in Table 81.

Table 81: Suggestions to improve soil condition

Sl. No.	Suggestions to improve soil condition	Rank
1.	Use of FYM, green manure and compost	I
2.	Inclusion of legumes in rice based cropping system	II
3.	Gradual replacement of chemical fertilizers	III
4.	Avoiding over grazing in summer	IV
5.	Efficient use of the irrigation water	VI
6.	Use of bio-fertilizers	V
7.	Fertiliser use on the basis of soil test	VII

The above suggestions are general in nature. The farmers are not aware of other packages. They insist on use of FYM and compost, which were the practices over past generations to keep soil rich. But after use of chemical fertilizers the status of soil has been changed. The farmers are in opinion that present soils of their fields do not respond as earlier. The suggestions further revealed that emphasis on use of FYM, green manure as well as supply of seeds of green manuring crops, inclusion of pulse in crop rotation should be encouraged to reduce dependency on chemical fertilizers. The summer grazing and water management practices in irrigated area need a great deal of planning to restore the fertility level of soil. The last suggestion, use of fertilizer on soil test also is genuine to prevent imbalance use of chemical fertilizers.

ii. Water

The coastal belt of Odisha enjoys the opportunity of irrigation water, and besides there is need to arrest run-off water through different water reservoirs for future use in farming and increasing water table of the area. The observed position of ground water table is not encouraging. The small farm holders offered the following suggestions to preserve water for use in agriculture in summer and incase of scanty rainfall.

Table 82: Suggestions to preserve ground water

Sl. No.	Suggestions to preserve water	Rank
1.	Conservation of run-off water in small ponds/tanks	I
2.	Recharging of ground water through contour bunding and terracing	II
3.	Use of sprinkler and drip irrigation	III
4.	Consumptive use of multi-source and multi-quality water	IV
5.	Adoption of recommended irrigation practices	V

The above suggestions are scientific recommendations to conserve water in the areas of investigation. The farmers of surveyed area feel that run-off water could be conserved erecting field bunds wherever possible. There are chemical based industries, which discharge chemicals to the rivers that is to be checked. To preserve water for scarce days, use of sprinkler and drip irrigation is the answer as suggested by the respondents.

iii. Forest

The forest area of the state is about 37.33% and the surveyed districts possess 11.74% of area under different kinds of forest. As per rule of the State Govt., steps are taken to restore the forest area. The sample farmers have suggested the following measures to restore the forest area within the locality.

Table 83: Suggestions to improve forest area

Sl. No.	Suggestions to preserve forest area	Rank
1.	Raising of forest	I
2.	Planting of trees in the vacant areas	II
3.	Prevention of forest fire	III
4.	Planting of non-timber forest trees	IV
5.	Raising of nursery for multipurpose tree species	V
6.	Preservation of forest through local committee	VI
7.	Adoption of participatory forest management system	VII
8.	Heavy penalty for offenders	VIII

The findings revealed that the respondents of the study area are conscious of the forest resources. They emphasized that community participation to manage the forest is of prime importance. There is need to create sensation among people about the forest. Raising of non-timber forest is another solution to reduce pressure on forest to meet the requirements of fuel and fodder of the people.

iv. Animal wealth

Animal wealth is an integral part of farming system. Odisha represents 9.16% to the national animal wealth. The surveyed districts have cattle population upto 19.25% of the state. The suggestions of the sample to preserve animal wealth were recorded as mentioned below.

Table 84: Suggestions to preserve animal wealth

Sl. No.	Suggestions to preserve animal wealth	Rank
1.	Preserving indigenous germ plasm	I
2.	Improvement of local breed	II
3.	Fodder cultivation	III
4.	Animal feed of local available feed materials	IV
5.	Regular animal health check-up	V

The suggestions contained in table above reveal emphasis on local indigenous breed, fodder cultivation and preparation of animal feeds on locally available materials.

11.2. Suggestions for profitable farming enterprise

The objective of sustainability of agriculture is not only to preserve ecosystem but also to provide higher income to sustain the farm and farming enterprise. This in turn will reduce manipulation of natural resource in development process. There have been a number of suggestions regarding profitability of agri-enterprise to supplement the concept of sustainability. The suggestions have been presented in Table 85.

Table 85: Suggestions regarding profitable farming enterprise

Sl. No.	Suggestions regarding profitable farming enterprise	Rank
1.	Diversified farming system	I
2.	Production as per consumer preference	III
3.	Value addition to farm products	IV
4.	Market link	II
5.	Adoption of cost effective technology	VII
6.	Provision of agro-service centers	VI
7.	Emphasis on use of renewable energy	VIII
8.	Training for capacity building	V
9.	Priority of benefit to poor	X
10.	Women empowerment	IX
11.	Timely dissemination of farm technology	XIII
12.	Involvement of SHGs	XI
13.	Technology support at doorstep	XII
14.	Importance on participatory decision making	XVII
15.	Involvement of micro-level institutions and organizations/ NGOs	XIV
16.	Development of local leadership	XV
17.	Operation of village committee for sustainable agriculture	XVI
18.	Partnership between public and private agencies	XXIII
19.	Use of graded technology	XVIII
20.	Alertness to disaster management	XXI
21.	Provision of seed bank, community nursery and fodder production	XIX
22.	Emphasis on honesty and integrity	XX
23.	Working with common consciousness	XXII

11.3. Suggestions for strengthening extension system

Further, the extension components in maintaining sustainability in agriculture is to be given due importance to bridge the gap between farmers, and Subject Matter Specialists (SMS) and scientists. To initiate the concept of sustainability and motivate the small farm operators in particular, a new approach in extension system is necessary. Out of the findings of the study, the following measures are suggested to strengthen the extension system to work towards sustainability of agriculture.

Table 86: Suggestions to strengthen extension system

Sl. No.	Suggestions to strengthen the extension system	Rank
1.	Organisation of sustainable committee at village level	I
2.	Regular functioning of committee	II
3.	Inviting experts for technological support	III
4.	Co-ordinating the line departments	IV
5.	Contact with outside agencies for suggestion and technical help	V
6.	Encouraging indigenous technical knowledge (ITK)	VI
7.	Following the procedure of proper planning	VII
8.	Regular evaluation in the progress of the sustainable activities.	VIII
9.	Encouraging innovative ideas	IX

The suggestions contained in above table like, organisation of sustainable committee on agriculture, regular functioning of committee and visit of experts on technological matters in suitable intervals would able to bring sustainability in agriculture. The other factors like, co-ordination with line departments, contact with outside agencies, use of ITK, proper planning, evaluation and inclusion of new ideas would help to achieve sustainability in agriculture to a great extent.

12

Conclusion

The study entitled "sustainability of small farms in coastal ecosystem" was conceived out of consideration of continuous degradation of natural resources-base consisting of soil, water, forest, maintain and animal wealth of the state. With broad objective of investigation into the status of sustainability of agriculture in the state, the specific analysis like socio-economic and farming related situations, perception about sustainability of agriculture in terms of ecological soundness, economic viability, social accessibility, humaneness and adaptability, development of sustainable agriculture index, adoption of sustainable measures, extension components and suggestions and strategy to maintain sustainability were taken into consideration. The organic farming and agri-business opportunities for sustainable agriculture were also studied.

The study was conducted in the state of Odisha covering the districts of Puri, Khurda, Ganjam and Balasore. Altogether four blocks each of one district and eight villages two from each block were selected for the purpose of the study. As much as 200 respondents who were within definition of small and marginal farmers were selected for interview. The respondents randomly so selected were personally interviewed by the author through an interview schedule developed on the basis of objectives stated earlier. The secured data were processed statistically to reveal the relevant findings as per the need of the study. From the relevant findings of the study the conclusions have been summarized below.

1. The present study carried out in coastal area of Odisha yielded suggestions and strategy for up keeping the status of agriculture in the state. The

strategy has been presented under two important sections mainly technological measures and policy with extension approach. The technological measures are soil, water, forest and animal wealth.

2. To improve the soil condition of the area of study the suggestions included use of FYM, green manuring, compost, inclusion of legumes in rice based cropping system, gradual replacement of chemical fertilizers, limiting over grazing of cattle in summer, use of bio-fertilizers, efficient use of irrigation water and fertilizer on the basis of soil test. The suggestions have been ranked as per mention of the sample.

3. To preserve water for the purpose of agriculture and other uses, the suggestions in order of merit were conservation of runoff water in small ponds and tanks, recharging ground water through contour bunding and terracing, use of sprinkler and drop irrigation on the basis of applicability, making consumptive use of water and adopting recommended procedures in irrigation for crop production.

4. Although forest area of the surveyed district is comparatively small to inland districts of the state, there is need to preserve and extend the forest area. The important suggestions include raising of forest on available land, planting trees in vacant places, prevention of forest fire, planting of non-timber forest trees to reduce pressure in forest and activating local communities to preserve forest through participatory approach and heavy penalty for offenders.

5. For preservation of animal wealth the people were in opinion that there is need for preserving indigenous germplasm, improving local breed, extending fodder cultivation manufacturing animal feed from locally available feed materials along with regular animal health checkup practices.

6. Policy and extension approach include means and ways to increase profitability of animal enterprise and adoption of extension methods for sensitizing rural people to work towards sustainable agriculture.

7. For making farming profitable and side by side maintaining sustainability, the suggestions include diversified farming system, market link, making production as per consumer preference and making value addition of the product. The other suggestions included are providing training for capacity building, providing agro-service centers, upgrading cost effective technology and emphasizing on use of renewable energy.

8. The study emphasized that to make agriculture sustainable attention should be paid to empower women, giving priority to poor people, involving SHGs

at ground level, programmes supported with providing technology at doorstep.

9. The other important suggestions include, timely dissemination of farm technology (IT), involvement of NGOs, developing local leadership, operation of village committee, selection of graded technology for farmers and emphasizing participatory decision making process.

10. Further suggestion like provision of seed bank, emphasis on honesty and integrity in community work, making people alert about disaster management and creating atmosphere to work with common consciousness. The study strongly believes that partnership between public and private agencies is to be developed to take care of agriculture and sustainability measures.

11. The findings revealed that majority of the farmers did not adopt the critical livestock management practices i.e. regular check up and treatment, balance feed, clear drinking water, cattle shed management and preparation of milk and milk product for family consumption and sale. That might be reducing the productivity of the animal wealth. Hence the farmers required to be trained and supplied with the critical inputs to exploit this opportunity of livestock rearing in mode of sustainable agribusiness in situ for better income and profit.

12. The small farm operators also posed gap in skill level (31.66 %) and educational level (30.00 %). The gap is more or less 30 % which reveals that there is scope for improvement for business activity through proper training and exposure in the field of suitable farm business development and management. Gap in favourable farming climate was to the tune of 30.66 % as revealed from the study. This is not so high and also keeps opportunity of managing with location specific technologies. Hence, the climate suitability is to be examined and proper measures may be taken addressing the climatic uncertainty to ensure a protection against any climatic aberrations for farm business activities.

13. Although majority(85%) of small farm operators opted not to use own source of finance at all and 10% of the farmers opted not to use credit at all; the majority of small farm operators opted meeting financial requirement from both own source and credit to run agribusiness as a sustainable venture. The agribusiness activities hence, need to be promoted both through technological and financial support. The human resource development in this context keeps priority.

14. Our sample is consisting of small and marginal farmers. The score analysis reveals that farmers are aware of non-use of chemical fertilizers in organic farming. The other components recognized by them are sound ecological environment, preservation of soil fertility, consumer satisfaction, and production of qualitative food grains. They are also aware of stability of yield factor, less cost of production, and adoption of minimum tillage. In spite of awareness the adoption of organic farming is far from expectation.

15. With advance in science and technology, there has been significant change in practices of organic farming. It is not new to our farmers. Due to over emphasis on inorganic farming and provision of incentives, the farmers of all land sizes are not willing to opt for organic farming. A good percentage of farmers have knowledge about organic farming practices like, growing of legume crops, green manure, keeping of farm animals and use of local varieties of crops. Very small percentage of farmers know about mixed farming, growing of herbs and shrubs to supplement organic contents of the soil. The need for popularization of organic farming practice is very much realized to boost the program.

16. The study shows that farmers have good knowledge about the benefits of organic farming. Majority of them know the benefits of organic farming in terms of better use of farm wastes, prevention of environmental degradation, reduction in cost of farming, improve in soil condition, good growth of crops, and improvement in environment. However, they have less knowledge about use of insoluble inputs, and nitrogen sufficiency in field. Although farmers of small and marginal groups have good amount of knowledge in organic farming, still all of them are in inorganic farming.

17. Problems in adoption of organic farming are many as expressed by the farmers in rural areas. These are lack of emphasis, absence of definite policy, lack of technical support, not sure of good yield, difficulty in identifying the organic area, labor intensive and lack of market demand. Many of these problems are not beyond the capacity to solve provided a sincere attempt is made.

18. The suggestions to increase adoption of organic farming are, creation of market for organic produce which will be definitely costly, incentives for organic farming by the Government, popularization of programme, special effort by Government, training, technical assistance, involvement of youth and formation of organic farmer groups.

19. The extension approach in the area need to be redesigned for bringing agricultural sustainability. The suggestions include organisation of sustainable committee at village level with regular operation of committee,

inviting experts for technological support and coordinating line departments for suggestions and inputs. However, suggestions also included keeping contact with outside agencies for technical help, encouraging indigenous technical knowledge (ITK), following the procedure of proper planning and undertaking regular evaluation of the project. However, encouraging innovative ideas of local farmers is a must not only to progress in agriculture, but also to keep agriculture moving without losing natural resource base.

Farmer's option also showed a gap of 48.33 % and 47.33 % in availability of technical support and interest of family members in farming respectively. The technical support as well as the interest of family members are considered to be the driving factors for business activities. Hence these factors having a higher gap should be improved for sustainable agribusiness of small farms. Awareness about recent technologies and the social status of the farmers also revealed gap of 39 % and 36 % respectively indicating that the farm business management require interest of farm family in agribusiness venture. The social status of small farm operators may not be the sole deciding factors as because the farm business is more an entrepreneurial action than cultivation only indicating a varied social status.

Bibliography

Agarwal, M.S. (1997). Planning for sustainable and regional development. Kurukshetra. Jan-Feb: 104-106.

Al Jalaud, A.A., Hussain, G., Yajima, M., Okada, K. and Matsumoto, N. (2002). Crop production and management in semiarid and arid environment. Proceedings of the 8th JIRCAs International Symposium, Tsukuda, Japan, 27-28 November, 2001. JIRCAs International symposium Series, 2002, No.10.105:116.

Alagesan, N. (1999). Farmers' knowledge on bio-fertilizers and bio-control agents. Journal of Extension Education 10(3): 2534-2536.

Ali, M.S.S. (1998). Developing sustainable agriculture programme putting the farmers knowledge. Journal of Social-Economic-Pertanion, Indonesia (1998), No. 4, 17-22.

Altieri, M.A. (1992). Agroecological foundations of alternative agriculture in California. Agriculture, Ecosystems and Environment, 39 (1 – 2) : 23-53.

Anderson, P. (2000). Insufficient investment. The Hindu Survey of Indian Agriculture, 21-23.

Atkinson, W., Watson, C.A., Pearce, B., Wood Warl, L., Wolfam Welh, J. and Nowa, C.K. (2002). Organic agriculture and GM crops. The BCPC conference: Pest and diseases. Proceedings of an International conference, Brighton U.K. 18-21 November – 2002, British Crop Protection Council; Farmham, U.K. 523-530.

Barad, A.V., Poshiya, V.K., Dhaliwal, G.S., Arora, R., Randhawa, N.S. and Dhawan, A.K. (1998). Planning technology for sustainable agriculture. Ecological, agriculture and sustainable development. Volume 2. Proceedings of International Conference on Ecological Agriculture Towards Sustainable Development, Chandigarh, India, 15-17, November: 419-429.

Basawarajaiah, V. (2001). Impact of Edira watershed development programme on farm families in Mahaboobnagar district of Andhra Pradesh. M.Sc (Ag.) Thesis Unpublished, Acharya N. G. Ranga Agricultural University, Rajendranagar, Hyderabad. 3.

Bautista, E.U. (2003). Mechanising rice agriculture in the Philippines. Proceedings of the Philippines Rice RDRE Scholar's Forum. Maligaya, Nleeva Ecija (Philippines). Phili Rice and PA BAR 2000, 316.

Bhavsar, V.M. (2002). Certified organic farming principle and practices, a course linking farmers and University students. Journal of Natural Resource and life Science Education, 2002, 31: 20-24.

Bhople, R.S. and Lakhdive, B.A. (1998). Use of neem seed extract for control of crop pests by farmers. Agricultural Extension Review, 10(3): 29-30.

Bhople, R.S., Shinde, P.S. and Dhule, S.S. (2001). Knowledge and adoption of bio-control pest management in cotton. Maharashtra Journal of Extension Education, 20: 18-21.

BIFADEC (1991). Sustainable Agriculture is the solution, but what is the problem? Occasional paper no.14, Agency for International Development, Washington, 69.

Borkar, M.M., Chothe, G.D. and Lanjewar, A.D. (2000). Characteristics of farmers influencing their knowledge about use of bio-fertilizers. Maharashtra Journal of Extension Education, 19: 130-131.

Brundtland, (1987). "World Commission of Environment and Development", developing the broad political concept of sustainable development, published its report, Our Common Future, April 1987: 10.

Bussche, P.F.V., Bussche, P.F.V. and Bussche, P.F.V. (2002). Sustainable development in Germany and Europe, the changing paradigm. Entwicklung – Landlicher –Rau, 36(4) : 17-19.

Cai, Y. and Smith, B. (1994). Sustainability in agriculture: a general review. Agric-ecosystem environ Amsterdam; NewYork: Enseeier, 193-June, 1994, 49(2): 299-307.

Chakraborty, T., Mallik, R.B., Dasgupta, M.K., Ghosh, D.C, Dasgupta, D., Majumdar, D.K., Chattopadhay, G.N., Ganguli, P.K., Munsi, P.S. and Bhattacharya, D. (1995). Concept of sustainability in agriculture under Indian context – major issues and resource recycling. Proceedings National symposium on sustainable agriculture in sub-humid zone, March 3-5, 2 : 12-19.

Conway, G.R. (1990). The challenge of sustainable agriculture. Environmental Education and Sustainable Development, 5: 185-186.

Cothe, G.D. and Borkar, M.M. (2000). Constraints faced by farmers in adoption of bio-fertilizers. Maharashtra Journal of Extension Education, 19 (4): 298-299.

Cox, J.W., Mc. Vicar, T.R., Reuter, D.J., Wang, H., Cape, J. and Fitzpatric, R.W. (2002). Assessing rainfed and irrigated farm performance using measures of Water Use Efficiency, Regional water and soil assessment for managing sustainable agriculture in China and Australia, 2002, 70-81.

Dabrowski, Z.T. (2000). The necessity of changes in the methodology of development and implementation of integrated pest management. Progress in Plant Protection, 40(1) : 334-342.

Dhillon, R.P. (2000). A study on farmers' awareness regarding agricultural pollution in Punjab. M.Sc. thesis, Punjab Agricultural University, Ludhiana, India.

Dunlap, R.E., Beus, C.E., Howell, R.E. and Wand, T. (1992). What is sustainable agriculture? An empirical examination of faculty and farmer definitions. Journal of Sustainable Agriculture 3(1): 5-39.

Eichhorn, H., Evcim, U., Dengirmencioglu, A., Demir, V., Yalcin, H. and Ozden, K. (2002). The role of agricultural Engineering in sustainable farming under regional and global challenges. Proceedings of 8th International Congress on Mechanisation and Energy in Agriculture, Kusadasi, Turkey-15-17 November, 2002, 1-4.

Elangovan, R. and Vasanthakumar, J. (1997). Perception of extension officials towards eco-friendly technologies. Journal of Extension Education 8(2): 1755-1758.

Fadini, M.M.M. and Lauzada, J.N.C. (2001). Environmental impacts of conventional agriculture. Infome agripecuario, 2001, 22(213): 24-29.

Flora, C.B. (2001). Access and control of resources. Lessons from SANREM CRSP. Agriculture and Human values, 2001, North Central Regional Centre for Rural Development, Department of Sociology, IOWA State Univ. 18(1) : 41-48.

Gelderman, U. and Kogel, K.H. (2002). Natured Concept. The new agriculture amidst ecology and economy of the gene. Journal of Agronomy and Crop Science, 2002, 188(6): 368-378.

Ghadei, K. (1996). Control of environmental pollution for sustainable agriculture as perceived by the farmers. M.Sc. (Ag.) Thesis, Orissa University of Agriculture and Technology, Bhubaneswar. 3.

Gips, T. (1986). What is Sustainable Agriculture? In Allen Pand van Dusen D. (Eds.), Global perspectives on Agroecology and sustainable agricultural systems. Proceedings of the 6th International Scientific Conference of the International Federation of Organic Agriculture Movements (Santa Cruz: Agroecology Programs).

Gowda, M.J.C. (1996). Sustainability of rice farming in different ecosystems. Ph.D. Thesis Unpubl., University of Agricultural Sciences, Bangalore.

Gowda, V., Govinda, L. M.T. and Chandrasekhar, S. V. (2000). Knowledge of farmers on dryland farming practices in groundnut cultivation. Journal of Extension Education 11(1): 2651-2655.

Haque, S.M.A., Haque, M.E. and Hossain, M.A. (1999). Farmer's knowledge on use of organic matter in crop production. Bangladesh Journal of Training and Development 12(1-2): 19-24.

Harrington, L.W. (1992). Measuring sustainability and alternatives. Journal of Farming Systems Research and Extension 3(1): 1-20.

Hasbi, H. (2001). Management of agriculture for organic coffee by using organic matter to improve the sustainability. Sustainable development in the context of globalization and locality: Challenges and options for network in South East Asia. Proceedings of the workshop, Bogor, Indonesia, 18-22 September 2000, Trompenlandwrit, Beiheft, 2001, 73: 136-149.

Hoff Gimenez, E. (2002). Measuring farmers agro-ecological resistance after Hurricane, Mitch in Nicaragva: a case study in participatory, sustainable land management impact monitoring, agriculture, ecosystem and environment, 2002. Department of environmental studies, 321 Nature sciences, 2, university of California, Santacruz, 1156 High Street, Santacruz, CA 95064, USA. 93(1-3): 87-105.

International Green Front Report (1998). Permaculture, pp:106-122.

Izac, A.M.N., Swift, M.J. and Dverak, K.A. (1992). Unpublished paper, CGIAR Social Science Meet, ISNAR, The Hague Aug. 1992. 2.

Jalali, B.L., Jalali, I., Dhaliwal, G.S., Arora, R., Randhawa, N.S. and Dhawan, A.K. (1998). Dynamics of integrated disease management in sustainable agriculture. Ecological agriculture and sustainable development: Volume 2. Proceedings of International Conference on Ecological Agriculture: Towards sustainable development, Chandigarh, India, 15-17 November: 355-365.

Jayasree, K. and Prasad, R.M. (1994). Analysis of dimensions of sustainable agriculture. Journal of Extension Education 5(1): 822-826.

Jirli, B.V. and Gangadharappa, N.R. (1997). Cognitive domain of vegetable growers with respect to integrated pest management. Mysore Journal of Agricultural Sciences, 31(1): 98-102.

Jirli, B.V. and Gangadharappa, N.R. (1997). Psychromotor domain of vegetable growers with respect to integrated pest management. Mysore Journal of Agriculture Sciences, 31(3): 271-275.

Jitsenguan, T. (2001). Sustainable agricultural systems for small scale farmers in Thailand: Implications for the environment. Food and fertilizer Technology Center 2001, Food and Fertilizer Technology Center for Asian and pacific region, Taipei, Taiwan, 11 (11): 509.

Jonson, B. (2002). Crop protection in integrated production of field vegetables in Sweden - the status of IPM. The BCPC conference, Pests and diseases, Proceedings of an International conference Brighton 18-21 November 2002, 1 (1 & 2): 471-478.

Kadam, J.R., Patil, V.G. and Hardikar, D.P. (2001). Knowledge and adoption of soil and water conservation practices in watershed development project. Maharashtra Journal of Extension Education, 20 (1): 138-140.

Kalaskar, A.P., Shinde, P.S. and Bhople, R.S. (1999). Correlates of adoption of integrated pest management technology by cotton growers. Maharashtra Journal of Extension Education, 18 (2) : 45-48.

Kalaskar, A.P., Shinde, P.S., Bhople, R.S. and Geete, M.H. (2001). Factors influencing knowledge of cotton growers about integrated pest management practices in cotton. Maharastra Journal of Extension Education 20 (2) : 117-119.

Kassa, H., Gibbon, D. and Singh, B.P. (2002). Livestock improved house hold food security and sustainability of Ethiopian small farms. Journal of Sustainable Agriculture, 21(2): 73-93.

Katar, S. (1995). The watershed management approach to sustainability of renewable common pool natural resources: lessons of India's experience. Research paper, Institute of Rural Management, Anand, 14 (5) :17.

Katyal, J.C. (1997). Relevance of integrated concept. The Hindu Survey of Indian Agriculture pp. 25-31.

Katyal, J.C. Das, S.K., Rao, U.M.B. and Reddy Y.V.R. (1992). Towards sustainable Agriculture, Intensive Agriculture, July-August, pp. 3-11.

Kerlinger, F.N. (1973). Foundation of behavioural research. Holt, Rhinehart and Winston Incorporation, New York, pp. 112.

Kern, M.S. and Geiss, J. (1999). Endosulfan (Thiodan – R) – a cornerstone of IPM/ ICM. Proceedings of the 5th International Conference on Pests in Agriculture. Part – 3, Montpellier, France, 7-9 December, 835.

Keufer, E. and Van Elson, T. (2002). Nature conservation advising for farmers. Survey amongst owners of organic farms in lower Saxony. Natur sehutz-Und-Landshafts-planung, 34(10): 293-399.

Khabeta, T.S. M. (2002). Farm size and soil loss: prospects for a sustainable agriculture in kwazulu – Watal. Agrckan, 41(2): 135-146.

Khan, M.S. (1999). Critical analysis of eco-friendly technologies in rice cultivation – A study in an adopted village – Kondubhotlapalem. M.Sc.(Ag.) Thesis Unpubl. Acharya N.G. Ranga Agricultural University, Rajendranagar, Hyderabad. 9.

Koocheki, A., Dhaliwal, G.S., Arora, R., Randhawa, N.S. and Dhawan, A.K. (1998). A quantifying approach for evaluating sustainable agriculture in Iran. Ecological agriculture and sustainable development: Volume 2. Proceedings of International conference on Ecological Agriculture: Towards sustainable development, Chandigarh, India, 15-17 November, 1997, 10 : 1-5.

Kumar, A.A., Babu, B. and Ramachandran, U. (1999). Attitude of farmers towards agro-forestry programme in Kerala. Indian Ajournal of Forestry, 22(1 & 2): 155-159.

Kutty, J.K. (1996). Analysis of the management for sustainable agriculture by the farmers of Kerala. Ph.D. Thesis Unpubl. Kerala Agricultural University, Thrissur. 13.

Lahiry, S.C. (1997). Sustainable economic development. Kurukshetra, Jan-Feb: p. 39-40.

Lakhan, S., Sinha, B.P. and Singh, L. (1999). Constraints in watershed development in Doon Valley. Indian Journal of Soil Conservation, 27(1): 74-77.

Lampkin, N. (1990). Organic Farming. Family Press Books. IPSWICH, UK.

Landell, M.N. (2002). Marketing forest environmental services who benefits. Gatekeeper series. Sustainable agriculture and Rural Livelihoods Programme. International Institute for Environment and Development (IIED): London, U.K. No.104 : 22.

Lawrence, A. (1994). The meaning of sustainability for extensionists: Indicators for assessing technological innovations. International workshop on alternative and cost-effective extension approaches for sustainable agriculture: Methodological Issues, TNAU, Coimbatore. 4.

Lefroy, E.C. and Hobbs, R.J. (1992). Ecological indicators for sustainable agriculture. Australian Journal of Soil and Water Conservation 5(4): 22-28.

Lehar, F. Von (1999). Development of man effective conservation farming system through participatory on farm research. American Journal of Alternative Agriculture. 1999, 14(C) : 98-102.

Liu,G., Xu, M., Rui, L., Walker, J., Mu, W., Liu, GB., Xu, Mx., Rui, L., Hu, Wy., and Mc Vicar, T.R. (2002). Assessment of a small catchment on the Loess plateau. Regional water and soil assessment for managing sustainable agriculture in China and Australia. (Eds) Mc Vicar, T.R., Rui, L., Walker, J., Fitzepatrick, R.W. and Changming, L., NWSUAF, Yangling, Shaanxi., 712100, China, pp. 139-144.

Lohani, S.N., Chhetri, K.K. (2001). Women in sustainable agriculture development and environment. Agriculture and Environment Communication issue. Kathmandu (Nepal), Ministry of Agriculture and Cooperatives, pp. 7-14.

Loker, W.M. (1994). Where is the beef? Incorporating cattle into sustainable agroforestry systems in the Amazon Basin. Agroforestry Systems, 25(3) : 227-241.

Manjunath, L., Balamati, A.M. and Biradar, N.D. (1998). A study on differential perception of attributes of selected innovations by farmers. Rural India, pp: 40-43.

Manjunatha, B.N., Lakshminarayan, M.T., Anand, T.N. and Kumar G.T.P. (1999). Factors affecting adoption of sustainable sugarcane farming practices – A Discriminant function analysis. Mysore Journal of Agricultural Sciences 33: 375-378.

Mbata, J.N. (2001). Land use practices in Lesotho: implication for sustainability in agricultural production. J-Sustain-agric. Binghamton, N.Y. Food Product Press, C-1990-2001, 18(2&3): 5-24

Mc Donald, A., Vasconalas, M.C. and Lozzia, C. (2001). Low input viticulture and entomology cregon Proceedings of 20BC – WPRs Working group "Integrated Control in Vitivulture" at ponte de lima, Portugal, 2-7 March 2001, Bulletin OLLB – SROP, 24(7): 173-181.

Mendoza, T.C. (1994). Incorporating sustainability in national food production and agricultural extension programmes in the Philippines. Farm Management notes, 17(6): 32-43.

Muthuraman, P. (1995). Adoption pattern of organic manures in rice based cropping systems. Journal of Extension Education 6(3):1229-1232.

Nadakarni, M.V. (1993). Sustainable Development: Concept and issues with special reference to agriculture. Indian Journal of Agricultural Economics 48(3): 516-525.

Nagabhushanam, K. (1997). Sustainability of agriculture in watershed environment – An analysis. Ph.D. Thesis Unpubl. TNAU, Coimbatore.

Nagdeve, D.P. (1999). Plant protection status of IPM-trained dry paddy farmers in Vidarbha region of Maharastra state. Ph.D. Thesis Unpubl. Acharya N.G. Ranga Agricultural University, Hyderabad.

Nirmala, L. and Ravichandran, V. (1995). Factors influencing knowledge and adoption of bio-fertilizers. Journal of Extension Education 6(2): 1183-1184.

Nirmala, L., Ravichandran, V. and Radhakrishnan, T. (1995). Information sources utilization influencing knowledge and adoption of bio-fertilizers. Journal of Extension Education, 6(1): 1061-1064.

Norman, S. and Warmark. C. (2003). Farm sustainability in the semi arid tropics – a case study of small farms in Nicaragua using sustainability indicators. Minor-Field-Studies-International Office. Swedish-University of Agricultural Sciences, 222 : 93.

Ohlander, L, Lagerburg, C. and Gertson, U. (1999). Vision for ecologically sound agriculture systems. Journal of sustainable agriculture (1999), 14 : 73-79.

Padhee, A.K. (1995). Sustainable pest management. Yojana, 39(8): 13-14.

Padmavathi, M. (1996). Role perception and role performance of Mitra Kisans in National Watershed Development Project in Rainfed Areas (NWDPRA) in Chittoor district of Andhra Pradesh. M.Sc.(Ag.) Thesis Unpubl. Acharya N.G. Ranga Agricultural University, Hyderabad. 2.

Panigrahi, R. and Sarangi, B. (1996). Role of alternative farming systems in the quest for sustainable agriculture. Journal of Extension Education (Orissa), 1 (1 & 2): 117-119.

Paolethi, M.G., Stinner, B.D. and Lorenzone, G.G. (1989). Agricultural Ecology and Environment, Elsenier Science Publishers, Amsterdam, The Netherlands. p.1.

Paroda, R. S. and Anderson, P. (2000). Insufficient investment. The Hindi Survey of Indian Agriculture,2000, 21-23.

Pimental, D., Thomas, W., Chilliney, I. M.O.W., Buttler, D., Reinmann, J. and Kenneth, B. (1989). Practices in Agricultural Ecology and Environment Ed. Poloetti M. Stinner, B.R. and Loren-zoin G.G. Elsever Science publishers B.V., Amsterdam, 3-24.

Pornthip, K. (1998). People's participation in sustainable forest resource management. A case study of community forestry project. Bon Nam Ra Am phoekalono Saton Province, Bangkok (Thailand) pp. 150.

Prasad, M.S. and Sundaraswamy, B. (2000). Attitude of farmers towards dryland agricultural technologies. Journal of Extension Education 11(1): 2666-2671.

Rahman, M.Z. and Mikani, H. (1999). Farmer's attitudes towards sustainable agricultural issues and environmental quality in a selected area of Bangladesh. American Journal of Alternative Agriculture 14(1): 22-29.

Rahman, M.Z., Mikani, H. and Rahman, M.M. (1999). Towards sustainable farming development: The attitude of farmers in a selected area of Shimane prefecture, Japan. Journal of Sustainable Agriculture 14(4): 19-33.

Rai, J.P.N., Vikas, S., Pratibha, S.V.,Arunachalam, A. and Khan, M.L. (2001). Environmental management through sustainable agriculture. In: Sustainable management of forests, Ed. Arunachalam, A. International Book Distributors, Dehra Dun, India, pp. 493-502.

Rai, M. (2005). Internalising synergies. The Hindu Survey of Indian Agriculture, p. 20-27.

Raj, R.K. and Satapathy, C. (2003). A study on extension and management issues in implementation of watershed development project. Ph.D. thesis submitted to Orissa University of Africulture & Technology, Bhubaneswar, p. 146.

Raju, K. A. and Radhakrishna, M.C. (2000). Constraints faced by farmers in betelvine cultivation. Maharastra journal of Extension Education 19: 308-311.

Ramamurthy, V.S. (2000). A study on Janmabhoomi programme in Chittoor district of Andhra Pradesh. M.Sc.(Ag.) Thesis Unpubl. Acharya N.G. Ranga Agricultural University, Hederabad.

Rambabu, P. (1997). Indigenous technologies in cropping systems – An analytical study in Guntur district of Andhra Pradesh. Ph.D. Thesis Unpubl. Acharya N.G. Ranga Agricultural University, Hyderabad.

Rauta, (1997). Sustainable agriculture in Pemania. Revista-Societaloi-National Romane-Pentrce-Steinta-Solului-Seria, IIIa. 1997. 31 : 3-10.

Reddy, M.R. (1996). Plant protection training needs of vegetable growing farmers in Ranga Reddy district of Andhra Pradesh. M.Sc.(Ag.) Thesis Unpubl. Acharya N.G. Ranga Agricultural University, Hyderabad.

Reetz, H.F. (1990). Sustainable innovations – A part of not apart from conventional agriculture. Better Crops, Spring pp. 6-10.

Reijntjes, C., Bertus, H. and Waters, B.A. (1992). Farming for the future, ILEM, Netherlands, p.2.

Resmy, C, Shivamurthy, and Mahadevaiah, G.S. (2000). Knowledge level of small and big farmers in sustainable practices of coconut and banana intercropping. Land Bank Journal 39(4): 39-43.

Sabarathnam, V.E. (2002). R/R/PRA for Agriculture. Rapid, Relaxed and Participatory Rural Appraisal (Participatory Learning and Action). For Research and Extension in Agriculture (For crops and livestock) Vamsamarath Publications Hyderabad, 2002, 398-417.

Sadangi, B. (1997). Farmers response towards sustainable agriculture. Seminar report on sustainable agriculture organized by CART (Dec. 28-29th) at Orissa University of Agriculture & Technology, Bhubaneswar.

Saha, A.C. (2001). In the hands of people: an Indian case of watershed development, 2001, International Institute for Environment and Development (IIED), London, U.K., 78.

Sathapathy, C. and Das, R.N. (1997). Farmers of Orissa in contrast of sustainable agriculture. Seminar report on sustainable agriculture organized by CART (Dec. 28-29th) at Orissa University of Agriculture & Technology, Bhubaneswar.

Saxena, K.K. and Singh, R.L. (2000). Adoption of organic farming practices by farmers of Malwa region. Maharashtra Journal of Extension Education 19 : 53-55.

Sharma, A.K. (2003). Organic Product, making competitive with 3N's nature + native + neighbour, Agrobios Newsletter, 1 (13): 20.

Sharma, A.K. (2003). Search of sustainability. Agrobios Newsletter, 1 (13): 23.

Shinde, P.S., Bhople, R.S. and Vaidya, V.R. (1997). Adoption of integrated pest management practices by cotton growers. Maharashtra Journal of Extension Education 16: 167-174.

Shrivastava, L.S. (1998). Agriculture in India by 2000 A.D. – Emerging opportunities. National Book News Review, Bombay, 14(4): 38-43.

Siddiq, E.A. (2000). Yawning productivity gaps. The Hindu Survey of Indian Agriculture, 39-44.

Singh, K.A. (1997). Attitude of farmers towards social forestry programme: an analysis. Journal of Research ANGRAU, 25(2): 27-30.

Singh, M.P. and Singh, D.B. (2004). Biodiversity why? A land of Hue and cry at present. Agrobios Newsletter, 2 (8) : 14-16.

Singh, R. (2004). Role of chemical pollutants in Agro-Ecosystems and Environment – Some considerations. Agrobios Newsletter, 2 (8) : 54-56.

Singh, R.B. and Kelayutham, M. (2001). Balancing crop and animal productivity and environment. Proceedings of Asian Agricultural Congress, Manila, Philippines 24-27 August 2000, 314.

Singh, R.P., Paroda, A.S., Joshi, N.L., Kathju, S. and Amalkar, K. (1999). Sustainable crop production strategies for management of arid and semi-arid ecosystems. Recent advances in management of arid ecosystem. Proceedings of a Symposium held in India, March 1997: 91-120.

Srinivasa, G., Dolli, S.S., Raveendra, M. and Iyengar, M.N.S. (1996). Socio-economic factors and their relation to adoption of improved sericultural practices. Indian Journal of Sericulture 35(1): 43-45.

Sriram, N. and Palaniswamy, A. (1999). Farmers extent of adoption of eco-friendly agricultural practices in cotton cultivation. Journal of Extension Education 10 (1) : 2318-2323.

Stockle, C.O., Panpendick, R.L., Saxton, K.E., Campbell, G.S. and Van-Evert, F.R. (1992). A frame work for evaluating sustainable agricultural production systems, Am. J. Altern. Agric., Greenbelt, Wallace Institute of Alternative agriculture, 9(1-2): 45-50.

Sudhakar, G., Lourduraj, A.C., Rangaswamy, A., Subbin, P. and Velayutham, A. (2002). Effect of vermicompost application on the soil properties, nutrient availability, uptake and yield of rice – A review. Agricultural Reviews, 23(2): 127-133.

Sumathi, P. and Alagesan, V. (1998). A correlation analysis between the characteristics of respondents with their awareness and adoption behaviour of IPM in groundnut. Journal of Extension Education 9 (3) : 2143-2144.

Swaminathan, M.S. (1991). Dimensions of sustainable agriculture. In : Extension Digest, MANAGE, Rajendranagar, Hyderabad 4(2): 3-4.

Swaminathan, M.S. (2000). For an Evergreen Revolution. The Hindu Survey of Indian Agriculture, pp. 9-15.

Swaminathan, M.S. (2005). Three – pronged strategy needed. The Hindu Survey of Indian Agriculture, pp. 8-15.

TAC/ CGIAR (1989). Sustainable Agricultural Production: implications for international agricultural research, Rome, FAO, p.xi.

Teng, P.S. (1999). Sustainable agriculture: Industry's challenges. Agro. Chemicals News, 22(4) : 5-19.

Tubene, S. and Harson, J. (2002). The whole sale produce auction. An alternative marketing strategy for small farms. Am-J-attern-agric. Greenbelt, M.D: Henry A. Wallace Institute for Alternative Agriculture. 17 (3 & 4) : 18-23.

Venkataramani, G. (1991). Sustainable agriculture: Intensive research will pay off. The Hindu Survey of Indian Agriculture: 17-19.

Verma, M.C. and Agarwal, M.S. (1997). Planning for sustainable and regional development. Kurukshetra, Jan-Feb.: 104-106.

Vijaya, A.K. (1996). Sustainability through natural farming. Intensive Agriculture. Nov-Dec.: 9-10.

Wuest, S.B., Mc. Cool, D.K., Miller, B.C. and Vesth, R.J. (1999). Globalisation of Pesticide Technology and meeting the needs of low impact sustainable agriculture. American Journal of Alternative Agriculture, 1997, 14(2) : 50-58.

Yadav, B.K. (2004). Environmental Implications of organic Farming, Agrobios Newsletter, 2 (8): 18-19.

Yadav, P. and Mani, D. (2000). Mitigating pollution through integrated nutrient management system. Bioved, 11(1-2): 75-77.

Yadav, R.S., Jha, P.K. and Pandey, S.S. (1996). Sustainable agriculture for pollution free production. Indian Farmers Times, 27 (7): 11-13.

Appendices

APPENDIX 1

Title of the Study : Sustainability of Small Farms in Coastal Ecosystem.

INTERVIEW SCHEDULE

(Exclusively for farmers)

We are making a study to know the sustainability of agriculture of your area. You are requested to offer your free, frank and unbiased response.

Sl. No. ....Date ....

1. Name of the respondent :
2. Address - Village

   Block

   Dist.

   Pin code

PART - I

Socio personal Information

1. Age at present (completed years) :

2. Educational attainment

   a) Illiterate

   b) Primary

   c) M.E. School

   d) High School

   e) College and above

3. Caste

   a) SC

   b) ST

   c) OBC

   d) Others

4. Type of family you reside in

   a) Nucleus

   b) Joint family

5. What is the composition of your family ?

   Male ....

   Female ....

   Childrren ....

   Total ....

6. Membership in formal organisation

   a) Cooperatives

   b) Farmer's club

   c) Radio forum

   d) Pani Panchayat

   e) SHG

f) Farmers Interest Group

g) Vana Surakhya Samiti (VSS)

h) Any other

7. Exposure to mass media

A) How frequently do you read newspaper ?

a) Everyday

b) Occasionally

c) Sometimes

d) Never

B) Do you possess radio ? Yes / No

If Yes what programs really you listen ?

a) News

b) Entertainment

c) Weather information

d) Agricultural programs

e) Any other

(Please check up programs from daily newspaper and enlist)

C) Do you possess TV ? Yes / No

If Yes, what programs do you view in TV. ?

a) News

b) Political views

c) Entertainment

d) Programme on agricultural and allied field

e) Sports

f) Any other

8. What is your major and minor occupation ?

Occupation	Major (>60%)	Minor (<60%)
a) Farming		
b) Business		
c) Wage earning		
d) Service		
a) Any other		

(Major occupation means the sources from which more than 60% of the income comes from)

9. Since how many years you are in farming ? ....

10. What is your land position ? :

a) Total land owned ....

b) Total land under cultivation ....

c) Total land leased out ....

d) Total land leased in ....

11. What are the crops do you normally grow with area ?

Crop	Kharif (ha)	Rabi (ha)
a) Paddy		
b) Pulse		
c) Oilseed		
d) Vegetable crops		
e) Cash crop		
f) Any other		

12. How much of your area is under irrigation ?

Area under irrigation : Kharif ______________ ha

Rabi ______________ ha

13. What are the sources of irrigation ?

a) Canals

b) LI Points

c) Tank (pond)

d) Well

e) Tube well

f) Pond

g) Any other

14. What farm machinaries do you possess ?

Name of the machinary	Nos
1. Tractor	
2. Power tiller	
3. Sprayer	
4. Sprinkler	
5. Any other specify	

15. Animal wealth

Name of the animal	Nos
1. Cow	
2. Bullock	
3. Buffaloes	
4. Small ruminants	
Goat	
Sheep	
5. Poultry / Duck	
6. Rabbit	
7. Any other	

16. Where do you sell your farm produce ?

a) Own village

b) Nearby market

c) Block HQ

d) Dist. Headquarter

e) Outside district

f) Outside state

g) Do not sell at all

17.What is your opinion regarding selling price of your farm produce ?

a) Quite remunerative

b) Remunerative at times

c) Not remunerative

d) Less concern

18. If farm produce is not remunerative, what may be the reason ?

a) High Cost of production

b) Less Market demand

c) Middlemen take more profit

d) Products lack Value addition

e) No scope for storage

f) Any other please mention

19. What is your view about farm produce ?

a) Just manageable

b) Little over family maintenance

c) Much more than family requirement

d) Only grow for marketing

e) Any other reason

20. Do you have livestock and fishery to add to your family income ? Yes / No

If Yes, what are they

Items	Number
Milch cow	
Buffalo	
Bullock	
Poultry	
Duck	
Fishery (Pond of what size)	

21. Do you use energy in agriculture ? Yes / No

If Yes, how much energy bill do you pay / month or season or year ?

22. How many of your family members are involved in farming ?

a) Directly ____________

b) Indirectly ____________

23. Do you look after your farming ? Yes / No

If Yes, What are your rsponsibilities ?

a) Only supervisory work

b) Supervision & Management

c) Direct work in the field

d) Supervision, management & field work

e) Any other

24. Have you ever gone on any training on agriculture & allied ? Yes / No

If Yes, Where

Institution	Period
a) KVK	
b) Agril. Dept.	
c) NGO	
d) Any ot	

25. Have you ever used knowledge that you gained after coming from training? Yes/ No

If Yes, what new ideas you have adopted after being trained ?

1.
2.
3.

If No, What are the reasons ?

1. Not fully convinced with training
2. Lack of resource to put knowledge gained in training into action
3. Training not related to your problem
4. Any other reason (specify)

PART-II

Organic farming for sustainable agriculture

A. Awareness about components of organic farming

Statements	Average Mean Score	Rank
1. Non –use of chemical fertilizer and pesticides		
2. Use of local resources		
3. Preservation of soil fertility		
4. Pollution free agriculture		
5. Production of qualitative food grains		
6. Adoption of minimum tillage		
7. Bring stability in yield of crops		
8. Cost effective/less cost involved		
9. Consumer satisfaction		
10. Sound ecological environment		

B. Knowledge about benefits of organic farming

Benefits	Mentions	Percentage
1. Improves physical properties of soil		
2. More use of insoluble inputs		
3. N-sufficiency in farm		
4. Control of weeds		
5. Control of disease and pest		
6. Environment improvement		
7. Crops of good growth		
8. Better physiological activities of crop		
9. Better retention of soil moisture		
10. Reduction in cost of inputs		
11. Better use of farm wastes		
12. Consumer preference for organic food		
13. Better preservation quality of produce		
14. Preventation of environment degradation		
15. Checking of soil erosion		
16. Better scope for animal rearing		

C. Knowledge about Organic Farming Practices

Practices	Know		Don't know	
	Frequency	%	Frequency	%
1. Mixed farming				
2. Growing of herbs and shrubs				
3. Keeping of farm animals				
4. Crop Rotation				
5. Use of local genes				
6. None use of chemicals for soil, pest and diseases				
7. Vermicompost				
8. Compost making				
9. Green manure				
10. Growing of leguminous crops				
11. Use of Azolla				

D. Problems in adopting Organic farming

Problems	Average	Rank
1. Insufficient local resources		
2. Not sure of good yield		
3. Labor intensive		
4. Lack of market demand		
5. Lack of technical support		
6. Lack of organized efforts for organic farming		
7. Absence of Govt. policy		
8. Difficult to indentify organic areas for farming		
9. Apprehension to lose incentives given for fertilizers etc.		
10. Lack of emphasis and movement for organic farming		

E. Suggestions of farmers for adoption of organic farming

Suggestion	Mentions	Percentage
1. Incentives for organic farming		
2. Creation of market for organic produce		
3. Training and extension support		
4. Formation of organic farmer groups		
5. Involvement of rural youths		
6. Special program by govt.		
7. Popularization of organic farming by extension agencies		

PART – III

(A) Factors of Sustainabuility of Agriculture

Sustainability in Agriculture is discussed at various levels. There may be many factors of sustainability. We have listed some. Please add if you feel other factors are there and rate them assigning ranks.

Sl. No	Factors	Rank
1.	Ecological soundness	
2.	Economic viability	
3.	Social accessibility	
4.	Humaneness	
5.	Adaptability	

(Rank as I, II, III, IV and V in order)

(B) Perception About Sustainabuility of Agriculture

Ecology is one of the important factors of Sustainability. Ecological Soundness as such consists of many ingredients. Please record your response against each statement in a five point scale as you feel appropriate.

SA – Strongly agree, A- Agree, UD- Undecided, DA- Disagree, SDA- Strongly Disagree

Humaneness

Sl.No.	Statement	SA	A	UD	DA	SDA
1.	The soil condition of an area with regards to , its structure & fertility, level of organic matter reveal the soundness of the ecology.					
2.	Availability of groundwater and its nearness to root zone of crops indicate good ecological status.					
3.	The maintenance of nutrients of soil in form of NPK and other minor elements helps to have sound ecology of an area.					
4.	Types of farming including crop rotation, land use pattern etc are the main indicators of ecological soundness leading to sustainability.					
5.	Use of organic fertilizers such as FYM, Compost and Green Manuring leads to good ecology and sustainability in agriculture.					
6.	Participatory decision making for use of common resources in village is another important out come of sustainability.					

Economic Viability

Sl.No.	Statement	SA	A	UD	DA	SDA
1.	Level of production and its stability are important factors of economic feasibility that leads to sustainability.					
2.	Conservation of natural sources is an important factor to have economically feasible agriculture.					
3.	Increase in profitability in the existing farming system leads to economic feasibility.					
4.	Sustainable in agriculture provides enough income to meet the requirement of family.					
5.	Livelihood management consisting of animal and crop component on a profitable manner indicates the economic feasibility leading to sustainability.					
6.	Cost of production should be less for more profit.					

Social Accessibility

Sl.No.	Statement	SA	A	UD	DA	SDA
1.	Sustainability in agriculture ensures accessibility of all families to consume village resources.					
2.	Availability and use of energy in Agriculture is an Important component of social accessibility leading to sustainability in agriculture.					
3.	To maintain sustainability adequate technical / technological support is necessary.					
4.	Availability of capital at local level indicates social accessibility for sustainable agriculture.					
5.	Sustainability of agriculture opens wide scope in marketing of farm produce.					
6.	Participatory decision making for use of common resources in village is another important out come of sustainability.					

Humaneness

Sl.No.	Statement	SA	A	UD	DA	SDA
1.	Humanness comes where respect for honesty and value system operates.					
2	It develops cordial relationship among different communities.					
3	Sustainability implies respect for honourable living.					
4	Sustainability can create climate of trustworthiness.					
5	Sustainability of agriculture creates opportunity for compassion.					
6	Social solidarity is outcome of the sustainability in agriculture.					

Adaptability

Sl.No.	Statement	SA	A	UD	DA	SDA
1.	Sustainability is achieved where villages tend towards progressiveness.					
2.	Sustainability emphasizes accommodation of Govt. policies and technological change.					
3.	Sustainability makes market demand widened.					
4.	Consumer preference is important to have sustainability in agriculture.					
5.	Possible conflicts arise in adoption of sustainability methods/ practices and measures.					
6.	Sustainability provide scope for disaster management.					

(C) Sub-Factors of Sustainability of Agriculture

(A) We have secured your honest response on five important parameters of sustainability of agriculture. Under factors there are sub factors. Please rank them as you feel appropriate.

VI -Very Important, I - Important, LI - Less Important

Ecological Soundness

Item	VI	I	LI
1. Soil			
2. Water			
3. Nutrient			
4. Types of farming			
5. Use of Bio-pesticides and fungicides			
6. Use of organic manures			

Economic Viability

Item	VI	I	LI
1. Production			
2. Profitability			
3. Resource conservation			
4. Adequate income			
5. Profitable animal husbandry			
6. Profitable crop husbandry			

Social Accessibility

Item	VI	I	LI
1. Accessibility to village resources			
2. Availability of energy for agriculture			
3. Technological support			
4. Availability of Capital			
5. Market opportunity			
6. Participatory decision making			

VI = Very Important I = Important LI = Less Important

Humaneness

Item	VI	I	LI
1. Respect for honourable living			
2. Cordial relationship			
3. Honourable living			
4. Climate of trustworthiness			
5. Feeling for others			
6. Social solidarity			

VI = Very Important I = Important LI = Less Important

Adaptability

Item	VI	I	LI
1. Accommodation for social change			
2. Accommodation for Govt. Policy and technological change			
3. Creation of market demand			
4. Preference of consumers			
5. Possible conflict in adoption of sustainability measures			
6. Scope for disaster management			

VI = Very Important I = Important LI = Less Important

(Please see that under each factor, the sub- factors are ranked as per their relative position as far as you judge them correct).

Sustainability of Agriculture is measured in different terms. We desire to know your valuable knowledge level and experiences about existing / operational variables involved in Sustainability.

(B) Awareness on Organic Farming

Statements	Response
1. Non –use of chemical fertilizer and pesticides	
2. Use of local resources	
3. Preservation of soil fertility	
4. Pollution free agriculture	
5. Production of qualitative food grains	
6. Adoption of minimum tillage	
7. Bring stability in yield of crops	
8. Cost effective/less cost involved	
9. Consumer satisfaction	
10. Sound ecological environment	

PART – IV

Adoption of Recommended Measures of Sustainable Agriculture

Ecological soundness

Q 1. In case you mark degradation of soil, what measures do you know that can preserve soil status?

a. Addition of more organic matter

b. Crop rotation

c. Balance use of chemical fertilizer

d. Any other please specify

Q 2. To preserve the ground water for agriculture what practices do you follow?

a. Check run off of water

b. Grow vegetation

c. Create irrigation facilities

d. Develop watershed projects

e. Any other (mention)

Q 3. Do you follow crop rotation to maintain sustainability of agriculture ?

Yes—— / No ——

If 'Yes', please state the rotation that you follow

a. Paddy – paddy

b. Paddy – vegetables

c. Paddy – groundnut

d. Paddy – pulses

e. Any other (specify)

Q 4. Do you know about bio-pesticides, bio-fungicides, bio-herbicides or bio-control agents and bio-fertilisers in agriculture ? Yes .... / No.........

If 'Yes', Please give their names :

Product	Name	Using	Not using
a. Biopesticides			
b. Biofungicides			
c. Bioherbicides			
d. Biocontrol agents			
e. Biofertilisers			

Q 5. There are many natural resources in your area like Soil, Water, Forest, Mountain, Animal etc. Do you know the conservation methods of these resources ? Yes / No

If Yes, What are they ?

Practices	Followed	Not followed
a.		
b.		
c.		
d.		
e.		

Economic viability

Q 1. Do you know the methods of increasing the productivity of your land? Yes —— No ——

If 'Yes', What are they ?

a. Use of HYV. Seeds
b. Balance use of fertilizers
c. Irrigation management
d. Adoption of Post Harvest Technology
e. Use of farm machinery
f. Any other

Q 2. Do you follow the different means of increasing profitability of your farm?
Yes —— / No.——

If 'Yes', What are they ?

---

a. Adoption of new technologies
b. Use of information
c. Contact with market
d. Method of value addition
e. Any other (Specify)

Q 3. Please state the methods/means you know to make animal and crop husbandry profitable?

Methods

Crop	Animal
a.	
b.	
c.	
d.	
e.	

Social accessibility

Q 1. Please give your reaction for the following statements .

Sl.No.	Statements	VI	I	LI
1.	All the families of your village have equal access to the common resources			
2.	Energy like Electricity/ diesel/ animal are available to all in the village			
3.	Villagers Receive adequate technological support			
4.	Capitals are available farming			
5.	Prevalence of participatory decision making in common interest arises			

VI = Very Important I = Important LI = Less Important

Humaneness

Q 1. Humaneness is important for living as well as sustainable farming system. Please record your reaction against each statement given below.

Sl. No.	Statements	Very much	Much	Little
1.	There is respect for honesty in society			
2.	Good relationship exists amongst the communities			
3	People respect much for honorable living			
4	There is a climate of trustworthiness			
5	Feelings for others is common in the village			
6.	Social solidarity being maintained in the village			

Adaptability

Q. Please respond to the following questions as you feel appropriate.

Sl.No.	Statements	VI	I	LI
1.	The villagers welcome change in agriculture			
2.	People want to accommodate Govt. Policy and technological change in farming			
3.	Market demand as farm produce is in increasing trend			
4.	Consumer preference is receiving importance while planning for the produce			
5.	There is conflict in adoption of methods to have sustainability			
6.	At present people know the technique of disaster management			

VI = Very Important I = Important LI = Less Important

PART V

Agribusiness opportunities for sustainable agriculture

A. Human factors and gap

Sl. No	Factors	Mean Score	GAP (%)
1.	Family goal		
2.	Skill level		
3.	Family labour use		
4.	Educational level		
5.	Interest of family members in farming		
6.	Social status		
7.	Favourable farming climate		
8.	Awareness about recent technologies		
9.	Information sources		
10.	Availability of technical support		

B. Physical aspects of farm development

Sl.No	Physical factors	Level of adoption of recommended measures			Mean Score
		Full	Partial	Not at all	
1. Crop					
i	Short duration				
ii	Medium duration				
iii	Long duration				
2. Cropping pattern					
i	Sole crop				
ii	Double crop				
iii	Crop rotation				
iv	Inter cropping				
v	Mixed cropping				
3. Maintenance of soil fertility					
i	Legume based rotation				
ii	Use of organic matter				
iii	Use of chemical fertiliser				
iv	Use of crop residue				
4. Irrigation					
i	All farm				
ii	All crops				
iii	All seasons				

C. Animal husbandry

Sl.No	Factors	Adoption	Non- Adoption
1.	Regular check up and treatment		
2.	Balance feed		
3.	Clear water for drinking		
4.	Cattle Shed management		
5.	Preparation of milk and milk product		

D. Possesion and use of farm machineries

Sl.No	Factors	Possessors	Non-possessors
1.	Tractor		
2.	Power tiller		
3.	Power sprayer		
4.	Power weeder		
5.	Harvester / reaper / sprinkler		
6.	Winnower / thresher		

E. Financial aspects

Sl.No	Financial sources	Full	Partial	Not at all
1.	Own resource			
2.	Credit			
3.	Own resource and credit			

PART- VI

(A) Extension and Communication Elements

I. For maintaining sustainability in agriculture we need cooperation of people in the locality. To have cooperation what kind of structure do you suggest at village level?

Social structure	Response
1. Committee / organization for sustainability	
2. To utilize SHG for the purpose	
3. To reactivate Mahila Samiti and Youth Club for the purpose	
4. To take help of Govt. Policy	
5. Any other.	

II. To ensure cooperation what kind of mechanism do you suggest to sustain on durable basis.

Item	Response
1. Sustainable committee to sit frequently	
2. Inviting experts to committee	
3. Organising training on sustainability	
4. Field trip to areas of sustainable agriculture	
5. Any other suggestion.	

III. Sustainability is a time taking process. Every individual in the committee has to participate in the process. Please indicate how to evaluate progress in the maintenance of sustainability.

Item	Response
1. Periodical visit by experts	
2. Analysis of situation by sustainable committee	
3. Evaluating level of production and productivity	
4. Measuring level of living (socio-economic status)	
5. Study of reaction of people.	

IV. For sustainability of Agriculture marketing of farm produce is very much important component. To provide remunerative price to producer what kind of marketing network you suggest.

	Item	Response
1	Fixation of price of produce by committee	
2	Opening of Rayat bazaar/ Krushak bazaar	
3	Any other	

(B) Suggestions

It is now well recognized that the farming is to be made sustainable for better living. There are many suggestions. But we need your honest suggestions.

1. Ecological soundness

What are your suggestions to maintain ecological soundness of the area ?

(Ecological soundness means soil, water, forest, mountain, animal and other natural resources)

a. Soil

b. Water

c. Forest

d. Mountain

e. Animal

2. Economical viability

The present day agriculture does not seem to be economically viable. Please suggest measures to be adopted to increase economic viability without disturbing ecological soundness.

1. Productivity in crop
2. Productivity in animal husbandry
3. Overall profitability of farm

3. Social accessibility

It is observed that all the families in a locality / village do not have equal accessibility to common resources. Please suggest how this can be ensured?.

1. Use of energy in agriculture
2. Technological support
3. Availability of input
4 Common decision to share the resources

4. Humaneness

The present day society suffers greatly in the term of value system. Please suggest means and ways to maintain humaneness among the village people.

1. To maintain honesty
2. Good relationship among the people
3. Creating climate of trustworthiness
4. To maintain social solidarity

5. Adaptability

Adaptability of technology leads to progressiveness. But many people do not adopt them. Please suggest how to ensure the adaptability of new technologies?

1. To accommodate technologies in farming system
2. To participate in marketing of farm produce
3. Ensuring consumer preference for better market

Preparedness to face disaster if any (Disaster means flood, cyclone drought etc).

APPENDIX II

Title of the study : Sustainability of small farms in coastal ecosystem.

INTERVIEW SCHEDULE

(For Experts only)

We are making a study to know the sustainability of agriculture of your area. You are requested to offer your free, frank and unbiased response.

Name: Date ....

PART – I

(A) FACTORS OF SUSTAINABUILITY OF AGRICULTURE

Sustainability in Agriculture is discussed at various levels. There may be many factors of sustainability. We have listed some. Please add if you feel other factors are there and rate them assigning ranks.

Sl. No	Factors	Rank
1.	Ecological soundness	
2.	Economic viability	
3.	Social Accessibility	
4.	Humaneness	
5.	Adaptability	

(Rank as I, II, III, IV and V in order)

(B) SUB FACTORS OF SUSTAINABILITY OF AGRICULTURE

We have secured your honest response on five important parameters of Sustainability of agriculture. Under factors there are sub factors. Please rank them as you feel appropriate.

VI -Very Important, I - Important, LI - Less Important

Ecological Soundness

Sl. No.	Item	VI	I	LI
1.	Soil			
2.	Water			
3.	Nutrient			
4.	Types of farming			
5.	Use of Bio-pesticides and fungicides			

VI = Very Important I = Important LI = Less Important

Economic Viability

Item	VI	I	LI
1. Production			
2. Profitability			
3. Resource conservation			
4. Adequate income			
5. Profitable animal husbandry			
6. Profitable crop husbandry			

Social accessibility

Item	VI	I	LI
1. Accessibility to village resources			
2. Availability of energy for agriculture			
3. Technological support			
4. Availability of Capital			
5. Market opportunity			
6. Participatory decision making			

Humaneness

Item	VI	I	LI
1. Respect for honourable living			
2. Cordial relationship			
3. Honourable living			
4. Climate of trustworthiness			
5. Feeling for others			
6. Social solidarity			

Adaptability

Item	VI	I	LI
1. Accommodation for social change			
2. Accommodation for Govt. Policy and technological change			
3 Creation of market demand			
4. Preference of consumers			
5. Possible conflict in adoption of sustainability measures			
6. Scope for disaster management			

VI = Very Important I = Important LI = Less Important

(Please see that under each factor, the sub- factors are ranked as per their relative position as far as you judge them correct.)

PART – II

SUGGESTIONS

It is now well recognized that the farming is to be made sustainable for better living. There are many suggestions. But we need your honest suggestions.

1. Ecological soundness

What are your suggestions to maintain ecological soundness of the area ?

(Ecological soundness means soil, water, forest, mountain, animal and other natural resources)

a. Soil

b. water

c. Forest

d. Mountain

e. Animal

2. Economical viability

The present day agriculture does not seem to be economically viable. Please suggest measures to be adopted to increase economic viability without disturbing ecological soundness.

a. Productivity in crop

b. Productivity in animal husbandry

c. Overall profitability of farm

3. Social accessibility

It is observed that all the families in a locality / village do not have equal accessibility to common resources. Please suggest how this can be ensured? .

a. Use of energy in agriculture

b. Technological support

c. Availability of input

d. Common decision to share the resources

4. Humaneness

The present day society suffers greatly in the term of value system. Please suggest means and ways to maintain humaneness among the village people.

a. To maintain honesty

b. Good relationship among the people

c. Creating climate of trustworthiness

d. To maintain social solidarity

5. Adaptability

Adaptability of technology leads to progressiveness. But many people do not adopt them. Please suggest how to ensure the adaptability of new technologies?

a. To accommodate technologies in farming system

b. To participate in marketing of farm produce

c. Ensuring consumer preference for better market

Preparedness to face disaster if any (Disaster means flood, cyclone drought etc).

PART-III

AWARENESS ON ORGANIC FARMING

	Statements	Response
1.	Non-uses of chemical fertilisers and pesticides	
2.	Use of local resources	
3.	Preservation of soil fertility	
4.	Pollution free agriculture	
5.	Production of qualitative food grains	
6.	Adoption of minimum tillage	
7.	Bringing stability in yield of crops	
8.	Cost effective/less cost involved	
9.	Consumer satisfaction	
10.	Sound ecological environment	

NIPA Publications on Agronomy

S.No.	Title	Author	ISBN	Year
1	A Handbook of Minerals,Crystals,Rocks and Ores	Alexander, P.O.	9788190723787	2009
2	A Handbook of Soil-Plant-Water-Fertilizer and Manure Analysis	Durai, M.V.	9789381450185	2014
3	Acid Soils: Their Chemistry and Management	Sarkar,A.K.	9789381450383	2013
4	Advances in Soil Borne Plant Diseases	Naik, Manjunath	9788189422813	2008
5	Agriculture and Waste Management for Sustainable Future	Sannigrahi, A. Kumar	9789380235530	2011
6	Agrometeorology: A Simplified Textbook	Kathiresan, G.	9789383305568	2015
7	Approaches for Incorporating Drought and Salinity Resistance in Crop Plants	Chopra,V.L. & R.S.Paroda	9789383305742	2015
8	Bioinoculants: A Step Towards Sustainable Agriculture	Gupta, R.P. & A.Kalia: et.al.	9788189422219	2007
9	Biotechnology in India: Initiatives and Accomplishments	Niladri Bag ed.	9789385516252	2016
10	Biotechnology of VA Mycorrhizza: Indian Scenario	Chandra, S. & Kehri	9788189422226	2006
11	Botanicals as Ecofriendly Pesticides	Mahulikar, P.P. & K.M.Chavan	9788189422578	2007
12	Breeding,Biotechnology and Seed Production of Field Crops	Roy, Bidhan	9789381450680	2013
13	Cereal Grains: Evaluation,Value Addition and Quality Management	Saxena, D.C.:Ed.	9789381450857	2013
14	Cereals: Processing and Nutritional Quality	Ram, Sewa and B.Mishra	9789380235073	2010
15	Climate Change and Agricultural Food Production	Kibria, Golam et.al.	9789381450512	2013
16	Climate Change and Chemicals: Environmental & Biological Aspects	Kibria, Golam et.al.	9789380235301	2010
17	Climate Change and Food Security	Datta, M.& N.P.Singh	9788189422387	2008
18	Climate Change and Natural Resources Management	Lenka & Lenka	9789381450673	2013
19	Climate Mitigation and Carbon Finance: Global Initiavities & Challenges	Sahoo, A.K.	9789381450024	2012
20	Computers in Agriculture: Fundamentals and Applications	Sharma & Bhatt	9789385516160	2015
21	Conducting An Effective and Successful Training Programme	Sontakki,Bharat S.	9789383305223	2015
22	Conservation Agriculture for Carbon Sequestration and Sustaining Soil Health	J.Somasundaram	9789383305322	2014
23	Droughts in Agricultural Production: Monitoring and Mgt.	G.G.S.N.Rao	9789385516009	2015
24	Encyclopedia of Agricultural Meteorology	Dhaliwal & Hundal	9789380235547	2011
25	Engineering and General Geology	Sawant, P.T.	9789380235516	2011
26	Enhancing Pulses Production: Technologies and Strategies	Gangwar, B. & Singh	9789381450772	2014
27	Farm Tools and Equipment for Agriculture	Surendra Singh	9789385516221	2016
28	Food and Nutritonal Security by Sustainable Agriculture: Methods to Attain and Sustain	Mishra, Kumar Bijesh	9789383305049	2014
29	Geographic Information System	Gurugnanam, B.	9788190851282	2009
30	Geoinformatics Applications in Agriculture	Singh,Anil Kumar	9788189422233	2008
31	Geospatial Technologies for Natural Resources Mgt.	Soam, & Sreekant	9789381450802	2013
32	GIS: Fundamentals,Applications and Implementations	Elangovan, K.	9788189422165	2006
33	Green Agriculture: Newer Technologies	Behera, K.K.	9789381450277	2012
34	Hill Agriculture: Economics and Sustainability	Sharma & Dwivedi	9789381450871	2014

S.No.	Title	Author	ISBN	Year
35	Illustrated Dictionary of Microbiology	Patel, Sunil	9788189422950	2008
36	Improving Productivity of Drylands by Sustainable Resource Utilization and Management	Dayal, Devi	9789385516191	2016
37	Integrated Farming System Practices: Challenges and Opportunities	Nanda, S.S.	9789385516207	2016
38	Modern Technologies for Sustainable Agriculture	Kumar, Sunil	9789381450611	2013
39	Nanotechnology in Soil Science and Plant Nutrition	Adhikari, Tapan	9789381450789	2013
40	Nitrogen Use Efficiency in Plants	Jain, Vanitha & P. Ananda Kumar	9789380235738	2011
41	Objective Agronomy	Vishwakarma,A.	9789380235127	2010
42	Organic Farming	Singh, A.K.	9789385516139	2015
43	Organic Farming Scope and Uses of Biofertilizers	Panwar, JDS & Amit Jain	9789385516184	2016
44	Organic Spices	Parthasarathy, V.A	9788189422844	2008
45	Pest Management and Residual Analysis in Horticultural Crops	Gulati Rachna & Beena Kumari	9789381450710	2013
46	Pesticides: Methods of Their Residues Estimation	Kumari, Beena & T.S.Kathpal	9789380235394	2010
47	Pigeonpea Hybrids and Their Production	A.N. Tikle, KB Saxena, and Yadava	9789383305957	2016
48	Plant-Microbe Interactions	Ramasamy & Kumar	9789383305834	2015
49	Practical Manual of Entomology	D.H.Lewin	9789380235905	2011
50	Precision Farming in Horticulture	Singh, Jitendar, S.K.Jain,L.K.Dashora	9789381450475	2013
51	Principles and Applications of Agricultural Meteorology	Patra, Alok Kumar	9789385516245	2016
52	Recent Advances in Biopesticides	Johri, Jayendra: eds.	9789380235219	2010
53	Remote Sensing Applications in Dryland Natural Resource Management	Gaur, Mahesh	9789381450321	2013
54	Samekit Krishi Pranali	Kumar, Sanjeev	9789381450154	2012
55	Seed Production of Field Crops	Mondal, S.S.	9788190723763	2009
56	Seed Science and Technology	Vanangamudi,K.	9789383305117	2014
57	Soil Conservation: Fully Revised and Updated 3rd ed.	Norman Hudson	9789383305971	2015
58	Soil Microbiology and Biochemistry	Hassan, G.Dar	9789380235134	2010
59	Soil Sampling and Methods of Analysis	Pal, Sushant	9789381450574	2013
60	Soil Science: An Elementary Textbook	Puri, A.N.	9789383305063	2015
61	Soil Testing and Analysis: Plant, Water and Pesticide Residues	Brajendra, Patiram	9788189422707	2007
62	Solving the Pulses Crisis	Singh & B.Gangwar	9789381450482	2013
63	Sustainable Agriculture: A Vision for Future	Desai, B.K. and B.T.Pujari	9788189422639	2007
64	Sustainable Environmental Science	Sahu, D.D.	9789381450208	2012
65	System Based Integrated Nutrient Management	Gangwar, B. & Singh	9789381450055	2012
66	Technologies for Sustainable Green Environment	Davamani, V.	9789381450420	2012
67	The Chemistry of Soil Constituents	D.J. Greenland	9789388516214	2016
68	The Chemistry of Soil Processes	D.J. Greenland	9789383305926	2015
69	Weed Science	Das, P.C.	9789383305261	2015
70	Women in Sustainable Agriculture	C.Satapathy & Mishra	9789383305056	2014

FSC
www.fsc.org
MIX
Papier | Fördert
gute Waldnutzung
FSC® C083411

Zeitfracht Medien GmbH
Ferdinand-Jühlke-Straße 7
99095 Erfurt, Deutschland
produktsicherheit@kolibri360.de